Master Mariner

Master
Capt. James Cook

Published in collaboration with
the Vancouver Museums and Planetarium Association

Mariner

and the Peoples of the Pacific

Daniel Conner and Lorraine Miller

Douglas & McIntyre, Vancouver

Copyright © 1978 and 1999 by Vancouver Museums and Planetarium Association

99 00 01 02 03   5 4 3 2 1

All rights reserved. No part of this book may be reproduced, stored in a retrieval system or transmitted in any form or by any means, without the prior written permission of the publisher or, in the case of photocopying or other reprographic copying, a licence from CANCOPY (Canadian Copyright Licensing Agency), Toronto, Ontario.

Douglas & McIntyre Ltd.
2323 Quebec Street, Suite 201
Vancouver, British Columbia V5T 4S7

Canadian Cataloguing in Publication Data
Conner, Daniel, 1948–
  Master mariner
    Bibliography: p.
    Includes index.

  ISBN 1-55054-723-2 (paperback)

  1. Cook, James, 1728–1779. I. Miller, Lorraine, 1948–
II. Title.
G246.C7C65   910'.92'4   C78-002118-5

The publisher gratefully acknowledges the role of the Vancouver Centennial Museum and Vancouver Maritime Museum in commissioning this work. The museums' generous aid, particularly that of sharing the photographic material assembled for the exhibition "Discovery 1778: Captain James Cook and the Peoples of the Pacific," made publication of this book possible.

Typesetting by The Typeworks, Mayne Island
Book and cover design by Nancy Legue-Grout
Printed and bound in Canada by Friesens
Printed on acid-free paper

The publisher gratefully acknowledges the support of the Canada Council for the Arts and of the British Columbia Ministry of Tourism, Small Business and Culture. The publisher also acknowledges the financial support of the Government of Canada through the Book Publishing Industry Development Program.

Foreword: James Cook
A Navigator for the Twenty-first Century

On the eve of the twenty-first century, it is appropriate that this book about James Cook be reissued and that the replica of his ship, HMS *Resolution* is visiting Vancouver in 1999. Appropriate for two reasons.

First, we could use a Cook to navigate the perils and opportunities of the new century because of his penchant for seeking truth at the expense of illusion, his willingness to adapt as change was required and his ability to look ahead with a reasoned imagination.

Second, Cook's third voyage to the Northwest Coast in 1778 was the watershed event in the history of this coast. Before Cook, the coast was an uncharted shoreline of thousands of miles, the ancestral home of the First Nations people who inhabited its shores and inlets, a coast rich in the resources of forest and sea. Cook's last voyage brought the report of these riches to the commercial interests of England and the United States, and that information led to the founding of the fur trade which, in a few years, led to the colonizing of the coast and the gradual decline of its Native population.

Only 222 years lie between Cook's visit to the Pacific Northwest Coast and today, a time span of just four or five generations. Standing on a foreshore of the coast he so ably charted, we can imagine that shore he saw and intuitively know that somehow it should be restored to something approaching the natural conditions he observed.

Cook's mission was exploration, a mission also open to us if we count exploration not so much on what is seen but how it is seen. For Cook, exploration was the art of the possible, not just to go farther than any man had gone but to go as far as a man could go. His was the perfection of the art, a model for us all as we set off to explore the science, politics, environment and human condition of the new century.

Sam McKinney
Research Assistant
Vancouver Maritime Museum

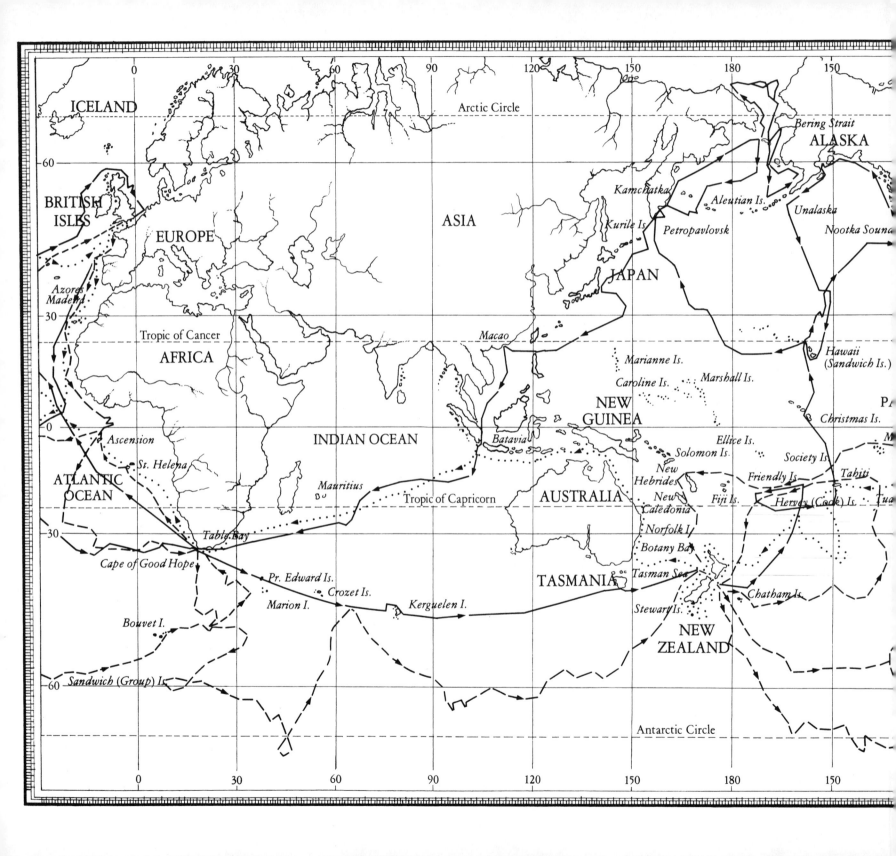

Contents

List of Illustrations viii
List of Engravings ix
Acknowledgements x
Introduction xi

PART I: PRELUDE 1
Chapter One: Apprenticeship 3
Chapter Two: First Voyage 8
Chapter Three: Second Voyage 13
Chapter Four: Planning the Third Voyage 20

PART II: THE THIRD PACIFIC VOYAGE 33
Chapter Five: Plymouth to Palmerston 35
Chapter Six: Tonga or the Friendly Islands 49
Chapter Seven: Tahiti and the Society Islands 66
Chapter Eight: Hawaii 1778 81
Chapter Nine: Nootka Sound 90
Chapter Ten: Prince William Sound and Cook Inlet 103
Chapter Eleven: Unalaska to the Arctic Sea 110
Chapter Twelve: Hawaii 1779 125
Chapter Thirteeen: Kamchatka and
 the Second Arctic Sweep 142
Conclusion: Kamchatka to Plymouth 157

Bibliography 162
Illustration Acknowledgements 164

THE THREE VOYAGES
Endeavour 1768-1771 ········
Resolution 1772-1775 — — —
Resolution 1776-1780 ———

List of Illustrations

Manuscript chart of the St. Lawrence by Cook   4
Cook's plan of the siege of Quebec   4
Cook's chart of Newfoundland   5
Engraving of Capt. Samuel Wallis's arrival in Tahiti, 1767   7
Cook's report on the transit of Venus from the South Pacific, 1769   9
The Endeavour River, New Zealand, with the *Endeavour* shown in repairs   12
Hadley's octant or quadrant   13
Larcum Kendall's First Chronometer, K1   14
Larcum Kendall's Third Chronometer, K3   15
Chart from *A Voyage towards the South Pole...*, 1777   16
Engraving from portrait of Joseph Banks   16
Map of Tartary, 1588   18
Map of Brobdingnag from *Gulliver's Travels*   19
Map of the Pacific Northwest pre-Cook (1768)   21
Capt. Charles Clerke, Royal Navy   22
Capt. John Gore, Royal Navy   23
Capt. William Bligh, Royal Navy   24
Capt. James King, Royal Navy   24
Capt. Edward Riou, Royal Navy   25
Capt. Nathaniel Portlock, Royal Navy   26
John Webber, R.A.   27
Omai   28
Engraving of Cook's murder in Hawaii, 1779   29

List of Engravings

Pl. 1 A View of Christmas Harbour, in Kerguelen's Land [Kerguelen Island] 39
Pl. 2 A Man of Van Diemen's Land [Tasmania] 42
Pl. 3 A Woman of Van Diemen's Land [Tasmania] 43
Pl. 4 An Opossum of Van Diemen's Land [Tasmania] 44
Pl. 5 The Inside of a Hippah [fortress], in New Zeeland 46
Pl. 6 A Man of Mangea [Mangaia, Cook Islands] 47
Pl. 7 A View at Anamooka [Nomuka, Tonga] 50
Pl. 8 The Reception of Captain Cook, in Hapaee [Tonga] 54
Pl. 9 A Boxing Match, in Hapaee [Tonga] 55
Pl. 10 A Night Dance by Men, in Hapaee [Tonga] 57
Pl. 11 A Night Dance by Women, in Hapaee [Tonga] 58
Pl. 12 Poulaho [Paulaho], King of the Friendly Islands 59
Pl. 13 Poulaho [Paulaho], King of the Friendly Islands, Drinking Kava 61
Pl. 14 A Flatooka, or Morai [sacred burial ground], in Tongataboo [Tongatapu] 62
Pl. 15 The Natche [Inasi], a Ceremony in Honour of the King's Son, in Tongataboo [Tongatapu] 64
Pl. 16 A Woman of Eaoo [Eua] 65
Pl. 17 A Human Sacrifice, in a Morai, in Otaheite [Tahiti] 71
Pl. 18 The Body of Tee, a Chief, as Preserved after Death in Otaheiti [Tahiti] 74
Pl. 19 A Young Woman of Otaheite [Tahiti], Bringing a Present 76
Pl. 20 A Dance in Otaheite [Tahiti] 77
Pl. 21 A Young Woman of Otaheite [Tahiti], Dancing 78
Pl. 22 A View of Huaheine [Huahine, Omai's home] 80
Pl. 23 An Inland View; In Atooi [Kauai] 83
Pl. 24 A Morai, in Atooi [Kauai] 84
Pl. 25 The Inside of the House, in the Morai, in Atooi [Kauai] 87
Pl. 26 A Man of Nootka Sound 92
Pl. 27 A View of the Habitations in Nootka Sound 94
Pl. 28 The Inside of a House in Nootka Sound 96
Pl. 29 A Woman of Nootka Sound 98
Pl. 30 Various Articles, at Nootka Sound 100
Pl. 31 A Sea Otter 102
Pl. 32 A View of Snug Corner Cove, in Prince William's Sound 104
Pl. 33 A Man of Prince William's Sound 107
Pl. 34 A Woman of Prince William's Sound. 108
Pl. 35 A Woman of Oonalashka [Unalaska] 111
Pl. 36 A Man of Oonalashka [Unalaska] 112
Pl. 37 Canoes of Oonalashka [Unalaska] 113
Pl. 38 The Tschuktschi [Chuckchi], and their Habitations 117
Pl. 39 Sea Horses [Walruses] 119
Pl. 40 Inhabitants of Norton Sound, and their Habitations 122
Pl. 41 Caps of the Natives of Oonalashka [Unalaska] 123
Pl. 42 Natives of Oonalashka [Unalaska], and their Habitations 124
Pl. 43 The Inside of a House, in Oonalashka [Unalaska] 127
Pl. 44 An Offering before Capt. Cook, in the Sandwich Islands [Hawaii] 128
Pl. 45 Tereoboo [Kalani'opu'u], King of Owyhee [Hawaii], bringing Presents to Capt. Cook 130
Pl. 46 A Man of the Sandwich [Hawaiian] Islands, with his Helmet 132
Pl. 47 A Man of the Sandwich [Hawaiian] Islands, Dancing 134
Pl. 48 A View of Karakakooa [Kealakekua], in Owyhee [Hawaii] 135
Pl. 49 A Canoe of the Sandwich [Hawaiian] Islands, the Rowers Masked 137
Pl. 50 A Young Woman of the Sandwich [Hawaiian] Islands 139
Pl. 51 Various Articles, at the Sandwich [Hawaiian] Islands 140
Pl. 52 A Man of the Sandwich [Hawaiian] Islands, in a Mask 141
Pl. 53 Summer and Winter Habitations, in Kamtschatka [Kamchatka] 144
Pl. 54 A Man of Kamtschatka [Kamchatka], Travelling in Winter 146
Pl. 55 The Inside of a Winter Habitation, in Kamtschatka [Kamchatka] 148
Pl. 56 A Man of Kamtschatka [Kamchatka] 149
Pl. 57 A Woman of Kamtschatka [Kamchatka] 150
Pl. 58 A Sledge of Kamtschatka [Kamchatka] 151
Pl. 59 A View of Bolcheretzkoi [Bolsheretsk], in Kamtschatka [Kamchatka] 153
Pl. 60 A White Bear 154
Pl. 61 A View of the Town and Harbour of St. Peter and St. Paul, in Kamtschatka [Kamchatka] 155

Acknowledgements

This publication was commissioned by the Vancouver Museums and Planetarium Association and grew out of the exhibition "DISCOVERY 1778: Captain James Cook and the Peoples of the Pacific," held at the Vancouver Centennial Museum in conjunction with the Vancouver Maritime Museum, in Vancouver, British Columbia, from 23 March to 6 September 1978.

"DISCOVERY 1778" was conceived as a bicentenary tribute to mark the arrival, on 29 March 1778, of Capt. James Cook at Nootka Sound on the west coast of Vancouver Island.

Both the exhibition and preparatory work for this publication were aided by grants from the Government of British Columbia, Captain Cook Bi-Centennial Committee, and the Government of Canada, National Museums of Canada, Museum Assistance Programmes.

We wish to thank the following institutions and their staffs for help in research and in obtaining illustrations, and to thank them for the courtesy with which they allowed these to be reproduced: The British Library, London, England; National Maritime Museum, Greenwich, London, England; Public Archives of Canada, Ottawa, Ontario; Public Record Office, London, England; Special Collections Division, The Library, University of British Columbia, Vancouver; Vancouver Museums and Planetarium Association, Vancouver.

In particular, we would like to thank Mrs. Anne Yandle, Head, and Miss Frances Woodward, Reference Librarian, both of the Special Collections Division, The Library, University of British Columbia, whose interest and encouragement greatly facilitated our research.

In addition, we would like to express our appreciation to Mrs. Lynn Maranda, Curator of Ethnology; Carol E. Mayer, Assistant Curator of Ethnology; Mr. Leonard G. McCann, Curator of Vancouver Maritime Museum, and Mr. Henry Tabbers, Photographer, all of the Vancouver Museums and Planetarium Association, and to Mr. Philip A. Yandle for their advice and support.

D.C. and L.M.M.

Introduction

All three Pacific voyages of Capt. James Cook have been remembered as great, even epic, journeys. Over a period of some ten years his explorations enlarged the known world far more greatly and more suddenly than in the thousand years before. He mapped New Zealand, crossed the Great Barrier Reef to chart the long eastern shoreline of Australia, discovered Hawaii, and examined the northwest coast of America, thus completing the outline map of the Pacific for the first time. No navigator before Cook had made voyages of such length or brought back so much accurate knowledge of such an immense extent of the earth.

Cook's achievements as seaman, navigator, surveyor, hydrographer and leader of men have attracted considerable notice from historians over the two hundred years since his death, but there is an important aspect to his voyages that has been neglected. The discovery of new lands meant the discovery of new peoples. Not only did Cook map the islands and coastline of the Pacific but also he, his officers and men, through their journals and sketches, recorded and commented upon the cultures they encountered. In addition to witnessing strange ceremonies, they noted these peoples' possessions, beliefs and ways of governing themselves, recorded their daily activities, learned a few words of their languages, and sketched their homes and countryside. These meticulous journals often provide our first account of the peoples of the Pacific, their behaviour and their thinking.

This study, as well as tracing Cook's nautical career, examines his encounters with native peoples by taking a detailed look at his Third Voyage. During this voyage Cook revisited peoples he was familiar with from his previous voyages and was able to bring to his observations the knowledge and insight gained by his earlier experiences. His journal, as well as those of his officers William Anderson, James King, David Samwell and Charles Clerke, are full of descriptions, impressions and evaluations of the Races of Man. The themes of the Noble Savage and the Depraved Savage, current in the Age of Enlightenment, emerge in the entries and in the judgements they make. They appear too in the sketches of native life and customs drawn by John Webber, whose work as official artist made this the most profusely illustrated of all Cook's voyages. The complete set of engravings from those sketches is reproduced here.

Both officers and artist were conscious of the need to keep their observations accurate and scientific, yet their journals and sketches unwittingly reveal the complex set of values and attitudes which coloured their view of the peoples they met. In the record of the Third Voyage we see just as much of the customs and habits of thought of the English as we do of the traditions of the Polynesians, Maoris, Tasmanians, Nootkans, Eskimos and Kamchadals.

Part I: Prelude

Chapter One: Apprenticeship

In September 1785, the last coat of arms granted for personal service to the British sovereign was posthumously awarded by King George III to Capt. James Cook of the Royal Navy, who had been killed in Hawaii six years earlier. Its heraldic description reads:

> Azure between two polar stars Or, a sphere on the plane of the meridian, showing the Pacific Ocean, his track thereon marked by red lines. And for crest, on a wreath of the colurs, is an arm imbowed, in the uniform of a captain of the Royal Navy. In the hand is a Jack on a staff proper. The arm is encircled by a wreath of palm and laurel.

The coat of arms carries two mottoes in Latin: the first, between the flag (the Jack) and the captain's arm, describes the range of Cook's achievement with *Circa Orbem* — Around the globe; the second, *Nil intentatum reliquit* — He left nothing unattempted — describes Cook himself.

Cook's professional education and the advancement of his career did not follow conventional lines. He wrote in 1775, "I have been allmost constantly at sea from my youth, and have dragged myself (with the assistance of a few good friends) through all the Stations, from a Prentice boy to a Commander." The greatness of his achievement is the more so because of his humble beginnings. He was born on 27 October 1728, in the Yorkshire village of Marton. The second son of a Scottish day-labourer and his English wife, he learned to read and write at the village school of Great Ayton before following his father as a labourer and then apprenticing as a grocer's assistant in the coastal village of Staithes.

In 1746, at the comparatively late age of eighteen, Cook signed as an apprentice seaman on a Whitby collier carrying coal between Newcastle and London. The coal trade along the dangerous, unlighted, unbuoyed English coastline was an exacting nursery for eighteenth-century seamen, for the ships were in constant peril of being forced on land by tides, winds and storms. Here, for nine years, Cook learned the practice and theory of seamanship, developing as second nature the ability to memorize a coast and its hazards. His training among the shoaling sands, tricky currents, ill-lit headlands and tidal ports of the North Sea coast would prove invaluable in years to come, for it developed his sense of danger so acutely that, on his Pacific voyages, his men would swear that he could smell unseen land before anyone else suspected its nearness.

Living conditions were appalling in the cargo-coasters of that time. Fighting gales, eating bad food, enduring bitter cold, living in cramped and bug-ridden quarters, Cook acquired the disregard for hardships that characterized his later career. Off-watch he applied himself to his books, sweating over the mathematics of navigation in the dim light of the crew's quarters. Between voyages he diligently continued his studies at the home of his employer, John

Walker of Whitby, with whom he had lodgings.

Cook's ability and competence were recognized and rewarded. By 1752 he was Mate of the collier *Friendship*. Three years later he was offered his own command and the chance of a share in a boat of his own. But just when his future in coastal shipping seemed settled and secure, Cook resigned. In 1755, at the age of twenty-seven, he signed as an ordinary seaman in the Royal Navy, then preparing to clash with France in the Seven Years War. Cook's voluntary recruitment was a most unusual step in an age when the lower deck was usually filled by the press-gang, and his motive has never been clearly understood. Some biographers have suggested ambition, others patriotism; recent research has even indicated that Cook might have enlisted to avoid a charge of smuggling. He gave no other explanation himself than that he had made up his mind "to try his fortune that way."

There were good opportunities in the eighteenth-century Navy for a practical seaman who knew his business. Cook quickly won the respect of his fellow crew members and recognition from his commanding officers. Within a month of his first posting to the 60-gun HMS *Eagle* on a blockade of the French coast, he was promoted to Master's Mate and had experienced his first action in the capture of a French East-Indiaman. Over the next two years he rose to Bo'sun and then to Master.

The Master in the eighteenth-century Royal Navy was a warrant officer with particular responsibility for navigation, surveying and the day-to-day running of the ship. Although he was not a commissioned officer, he was allowed to live and eat on board with officers of the rank of Lieutenant and above.

The manner of Cook's promotion to Master provides an early indication of the high regard in which his qualities were held. During the summer of 1757 the Captain of the *Eagle*, Sir Hugh Palliser, received a letter from the Member of Parliament for Scarborough urging the young Bo'sun's promotion to a commission, on the recommendation of his old employer, John Walker. The regulations which were in force at that time, however, required candidates for lieutenancies to have served on one of His Majesty's ships for at least six years. Palliser replied, therefore, that Cook "Had been too short a time in the service for a commission," but added "that a Master's warrant might be given him, by which he would be raised to a station that he was well qualified to discharge with ability and credit." His new responsibilities brought Cook in close contact with Captain Palliser. The two men developed a rapid respect for each other's ability. Palliser, who was later to be Governor of Newfoundland and a Lord of the Admiralty, had great influence on Cook's career, both as a firm friend and a powerful patron.

In 1757, at the age of twenty-nine, Cook was transferred as Master to the 64-gun *Pembroke* under Capt. John Simcoe, father of the future Lieutenant-Governor of Upper Canada, John Graves Simcoe. Captain Simcoe had an important influence on

Cook's life, encouraging and helping him in the study of astronomy and mathematics, and sharing his growing interest in the techniques of surveying.

During 1758 the *Pembroke* took part in the siege of the great French fortress of Louisbourg, which commanded the approach to the St. Lawrence River. Here, Cook came in contact with the noted military engineer, later to be Surveyor-General of Québec, Samuel Holland, under whose influence he began his vital training in hydrographic survey. Holland afterwards recalled in a letter to Lieutenant-Governor Simcoe that Cook had told him in 1776, when talking over the great surveying and mapping achievements of his first two voyages which produced the first outline map of the Pacific Ocean, that "the several improvements and instructions he had received on board the *Pembroke* had been the sole foundation of the services he had been enabled to perform."

As General Wolfe's fleet of troopships, with their convoying battleships, made their way slowly up the treacherous reaches of the St. Lawrence River towards Quebec, capital of French Canada, great demands were made on Cook's developing skill as a surveyor. The great obstacle was the river proper from the Saguenay to Québec, a 120-mile maze of shoals and rocky reefs from which all buoys and navigational marking aids had been removed by the French. Cook and the Masters of other ships were given the dangerous but essential task of checking captured enemy charts and marking a safe channel for the British fleet. The tedious and exacting work went on for months and because soundings had to be taken within artillery range of the city's massive fortifications, there was a constant danger of death or capture. Much, therefore, had to be done—with added difficulty—at night.

Cook was in his element. Distinguishing himself by the painstaking accuracy of his charting, he became much consulted; Wolfe himself sought his advice. Finally, on 26 June 1759, the British fleet under Admiral Saunders—22 warships and 119 merchantmen—negotiated the Traverse, the notorious passage from the north shore of the St. Lawrence at Cap Tourmente to the south shore between Ile Madame and the eastern tip of Ile d'Orléans. The troops were landed on the Ile d'Orléans the following day.

After the siege of Québec, Cook transferred as Master to the *Northumberland*, flagship of Admiral Lord Colville, who had succeeded Admiral Sir Charles Saunders as the commander-in-chief. In the summer of 1760 and 1761 the *Northumberland* operated in the St. Lawrence River and Gulf. On Colville's personal orders Cook carried out a complete survey of the St. Lawrence from Montréal to the Atlantic so that new charts could be made. His work was so thorough that many of his charts remained in use for more than a century. On 19 January 1761 Colville awarded Cook fifty pounds "in consideration of his indefatigable industry in making himself master of the pilotage of the river St. Lawrence." Again, on 30 December 1762, when forwarding Cook's surveys to the Admiralty, Colville informed their Lordships "that from my experience of Mr. Cook's genius and capacity I think him well qualified for the work he had performed and of greater undertakings of the same kind."

During the summer of 1762 the *Northumberland* operated on the south and east coasts of Newfoundland, taking part in the recapture of St. Johns from the French in September. In October 1762 she returned to England, where the ship's company, including Cook, were paid off. His skill and competence as a surveyor soon brought him back to the Maritimes, where from 1763 to 1767 he was engaged in a detailed charting of the Newfoundland coast—a task of vital importance to the British fishing fleet, as was well recognized by Governor Graves of Newfoundland, whose admiration Cook had quickly gained.

In 1764 Cook's old captain, Sir Hugh Palliser, became Governor of Newfoundland. On his recommendation, Cook at last received his first command, HMS *Grenville*, a ten-year-old, 70-ton schooner built in Massachusetts. Although still without a commission, Cook, as Surveyor of Newfoundland, was paid ten shillings a day, the same allowance given to the commander of a squadron.

Under Palliser's orders Cook extended his surveys still farther, adding Labrador to the area of Nova Scotia and Newfoundland in which he had worked. No previous hydrographic survey in these waters equalled Cook's work. His charts of some two thousand miles of practically unknown shore, deeply indented and notorious for bad weather, where obser-

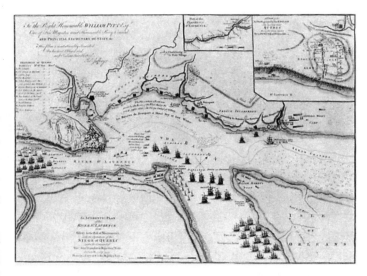

Cook's plan of the defences of the city of Quebec and its siege by Wolfe, 1759.

vations were often interrupted by summer fog and storms, were the first that could be trusted by seamen. Published in the *North American Pilot* in 1775, they became a standard navigational reference for over a century.

Cook's obvious talent for survey work brought him to the close attention of the Admiralty. His name also came before the scientific world. His report of an eclipse of the sun off the Newfoundland coast in 1766, and the calculation of the correct longitude from his very accurate observations, were published by the Royal Philosophical Society.

Cook at the age of forty, though still without a commission, had established a reputation for many kinds of ability. He was now to be singled out by the

Chart of Newfoundland.
From *The North-American Pilot*...by James Cook and Michael Lane, London, 1775.

Admiralty for "greater undertakings" which would test him to the utmost. When he returned to North America eleven years later he had become the greatest explorer and navigator of the age.

Chapter Two: First Voyage

In 1768 Cook, quite unknown outside a small circle of professional men, was promoted to Lieutenant and given charge of an expedition to the South Seas. His selection for the command was undoubtedly due to the influence of his patron and friend Sir Hugh Palliser, now on the Navy Board, who had long recognized Cook's high character and technical competence.

It required a sailor of exceptional talents to master the challenge of the Pacific in the eighteenth century. Its size and grandeur still challenge comprehension. Stretching ten thousand miles from China to the Americas and nine thousand miles between polar seas, it has large areas that are subject to almost perpetual calms, some to unceasing single-direction winds, others to typhoons and hurricanes.

This stupendous ocean, first seen by Europeans in 1513, was still largely unexplored in the middle of the eighteenth century. There was just a squiggle on the map to represent Abel Tasman's observation of western New Zealand in 1642; a blob to show Tasman's discovery of Tasmania; the simplest sketch outlines of Spanish and Dutch sightings of New Guinea and the north, south and west coasts of Australia.

Previous exploration, apart from its haphazard nature, had often been hampered by winds which severely limited the courses ships could steer, and by the persistent difficulty of accurately determining longitude: the distance east or west from a fixed point, which for English sailors was the Royal Observatory at Greenwich. Moreover, disease—often fatal—amongst seamen was made inevitable by poor victualling and ignorance of nutrition. As a result, large gaps had been left in outlining even those Pacific coasts that were known. Of the relatively few discovered islands, many had been lost again soon after being found. In particular, no one knew whether there was a navigable passage linking the northern Pacific with the Atlantic—the fabled Strait of Anian or Northwest Passage; no one had ever with certainty seen the east coast of Australia, and it was not clear whether there was a strait between Australia and New Guinea. Above all, no one knew whether the southern hemisphere contained ocean or a great land mass.

Many eighteenth-century geographers held to a very old belief that in the southern half of the world there lay a vast continent, the size of Europe and Asia combined, that was expected to bring great riches and power to the nation which discovered it. To classical geographers, the idea of a southern continent had seemed necessary to balance the weight of the land masses of the northern hemisphere. It was called "Terra Australis Incognita"—"The Unknown Land of the South"—and was thought to cover the whole of the southern surface of the earth. In the latter part of the Middle Ages a version of Marco Polo's travels in the thirteenth century, distorted by recopying and additions, claimed that a voyage south from Java would reveal "a golden province to which come few foreigners because of the inhumanity of the people." From the sixteenth century onwards, rumours and reports from voyages in the Pacific seemed to suggest that the great southern land was much more than a traveller's fantasy.

After the Seven Years War, continuing political and economic rivalry between the great sea powers—Britain, France and Spain—encouraged the sending of expeditions to discover this supposed continent of riches. Such were the secret instructions given by Britain to Commodore Byron on his world voyage from 1764 to 1766 and to Capt. Samuel Wallis on his circumnavigation from 1766 to 1768. In June 1767 Wallis, rediscovering Tahiti after its first sighting (some one hundred fifty years earlier) by the Portuguese navigator Pedro de Quiros, heightened hopes by his report that to the south he had seen mountains rising from the sea. "We supposed," he wrote, that "we saw the long wished for Southern Continent, which has never before been seen by Europeans." It was an illusion, perhaps a line of dark clouds against the setting sun, but the search now seemed more worthwhile than ever.

The scientific enthusiasm of the time provided a convenient excuse for the Admiralty to organize a secret voyage of exploration. In 1769 the path of the planet Venus would bring it between the sun and earth, an event that would not recur for another century and which was of great importance for astronomy. Observation of the transit from different parts of the northern and southern hemispheres, it was hoped, would enable scientists to calculate the distance of earth from the sun and thus provide a

Capt. Samuel Wallis's arrival in "Otaheite," June 1767.
Engraving by J. Hall from *An Account of the Voyages...for making Discoveries in the Southern Hemisphere*, ed. by John Hawkesworth, vol. I, 1773.

means of measuring the size of the solar system. Preparations were made by eight different nations for 150 observers to view the transit in places as far apart as northern Norway, Siberia, Hudson's Bay, California and Peking—a very early example of international scientific co-operation. The discovery by Wallis of Tahiti offered the chance to observe the transit from the South Pacific, and the Admiralty, seizing the opportunity, agreed to provide a ship for the purpose.

Despite Cook's minor rank, his skill as an astronomer and surveyor made him an ideal choice for the Admiralty's intentions. As well as being appointed official observer of the transit, he was given secret orders. After supervising operations on Tahiti he was to head south for some fifteen hundred miles, where, his instructions informed him, "there is reason to believe a continent, or land of great extent may be found." On finding this land he was to examine it in detail, establish "friendship and alliance" with the inhabitants, and "with their consent," possess the country in the name of the King. If no continent were to be found he was to turn west and investigate New Zealand, which many thought might be the northern tip of a greater land mass. He was then to return to England by whichever route he thought best.

The origins of the vessel that Cook chose for the voyage were as humble as his own, but it was selected with much thought. It was the plain but sturdy kind of Whitby collier in which he had first gone to sea. On 21 March 1768 the Navy Board recommended to the Lords Commissioners of the Admiralty to "make choice of a cat-built vessel," as the broad, flat-bottomed vessels were described, for the voyage to the Pacific, "which in their kind are roomy and will afford the advantage of stowing and carrying a large quantity of provisions so necessary on such voyages." Of three such vessels then in the Thames, the officers at Deptford yard reported most favourably on the Whitby-built collier the *Earl of Pembroke,* 308 tons, 106 feet long overall and about a third as wide, less than four years old, and with a running speed of seven to eight knots.

This homely vessel, with its broad, square stern and wide, bluff bow, was renamed the *Endeavour*. She was ideally suited for her task. She could operate at extreme range for long periods because she was roomy enough to carry adequate provisions and yet small enough to be taken ashore for repairs. Despite her large capacity, she was shallow and able to manoeuvre in tidal waters—a valuable advantage in exploring unknown coasts.

The *Endeavour* was bought out of the coal trade by the Admiralty for £2,800, a sum that was almost tripled in fitting her out for the voyage to the Pacific. This was no ordinary provisioning. The *Endeavour* was transformed into a floating laboratory, for as well as a voyage of exploration the expedition was to be the first major voyage of scientific discovery. Crowded on board with the officers and crew of eighty-six was a party of scientists and illustrators led by the wealthy young naturalist and patron of science Joseph Banks, future president of the Royal Society and founder of Kew Gardens. Banks had chosen to sail with Cook partly in place of making the fashionable "grand tour" of European capitals that completed the education of young gentlemen; the other part of his motive was to further scientific knowledge. At the age of twenty-five, therefore, he was making a tour of the world, having persuaded the Admiralty to allow him to join the voyage to Tahiti.

At his own expense Banks brought with him naturalists and artists to record the flora, fauna and sealife, the peoples they met and their ways of living.

The party included Dr. Daniel Solander, naturalist at the British Museum; Alexander Buchan, landscape and figure artist; Sydney Parkinson, botanical draughtsman; and Herman Sporing, naturalist. Between them, they gathered a wealth of new information about the peoples and lands of the South Pacific. They found animals, plants and minerals never before seen by Europeans, and brought back to England specimens, pictures and descriptions—including thirteen hundred new flowers—which immediately advanced scientific knowledge. Measurements were made of the earth's magnetism and gravity, the sea's temperature and salinity, the sea's tides and currents, the winds and the weather, the temperature and the pressure of the atmosphere. These able and gifted men must also have exerted an important influence on Cook, extending his interests, sharpening his intellect and deepening his sympathies for the native people whom he met on the voyage.

Cook's first Pacific adventure lasted three years. While the scientists catalogued their specimens and wrote about them, his talents as a seaman, navigator and hydrographer made their own spectacular contribution to men's knowledge of the shape of the world. After leaving Tahiti, Cook systematically explored both the east and west sides of the South Pacific, gradually narrowing the area where a new continent might lie. He charted some two thousand miles of New Zealand coastline, giving shape to a shadow which Tasman had only hinted at and proving conclusively that it consisted of two separate islands and therefore could not be part of a larger land mass. He was the first to navigate the Great Barrier Reef, although not without running afoul of its jagged coral, which grounded the *Endeavour* and made a gaping hole in her hull. He was the first to survey and chart the long east coast of Australia, the only great stretch of land still left unexplored in the temperate world. His voyage gave both Australia and New Zealand to the Crown. Moreover, his charts meant that these new lands could be readily found again.

Account and diagrams of the transit of Venus 3 June 1769 by James Cook.
Published in the *Philosophical Transactions* of the Royal Society, vol. 61, 1771.

While all Europe thrilled to his discoveries, Cook was modest about his achievement. He wrote to the Admiralty on his return, "I flatter myself that the discoveries we have made, tho' not great, will apologize for the length of the voyage." He could blend modesty, however, with a practical realization of the importance of what he had accomplished. As he wrote to his old employer and friend, John Walker of Whitby, comparing his contribution with the scientific catalogue brought back by Banks, "I however have made no very great Discoveries, yet I have explor'd more of the Great South Sea than all that have gone before me so much that little remains now to be done to have a thorough knowledge of that part of the Globe." He was already anticipating another voyage to obtain that "thorough knowledge."

Chapter Three: Second Voyage

Although Cook's first voyage did much to dispel belief in the myth of the southern continent, some geographers were unconvinced. In 1772 therefore, Cook, now Commander, set out to settle the question once and for all by searching not merely the South Pacific but all the oceans of the southern world.

The *Endeavour* was not available for this voyage, since, after her return to England in 1771, she had been sent to the Falkland Islands as a storeship. As Cook's experience on the Great Barrier Reef had shown him the risks of exploring with only one ship, two replacements were bought: the *Resolution* and the *Adventure*, both, like the *Endeavour*, east coast colliers. The *Adventure,* 368 tons, sailed under Capt. Tobias Furneaux, with five officers and seventy-five men. Cook sailed in the *Resolution,* 462 tons, nearly 111 feet long and about 35½ feet maximum width, with the collier's wide round-cheeked bow and roomy hull providing unusual space for stores, men and livestock. As a collier, she had carried a crew of perhaps twenty. She now had to accommodate over one hundred men and officers.

As with the *Endeavour,* her flat bottom and shallow draft enabled Cook to put in close to shore to chart coasts and to land easily. He also knew she would not overbalance if she went aground on a submerged rock, reef or sandbank. She was stable and sailed easily. By the standards of the day her appearance was far from elegant, but Cook saw her qualities. He wrote of her that she "was the properest ship for the service she is intended for of any I ever saw." She proved to be "one of the great, one of the superb ships of history." Cook historian J. C. Beaglehole has said of her that "of all the ships of the past, could she by magic be recreated and made immortal one would gaze on her with something like reverence."

This time Banks did not accompany the expedition. The *Resolution* could not be safely adapted to his grandiose plans for housing his large entourage in the comfort that he expected. Unable to contemplate the discomforts of a long voyage, which Cook and his officers took for granted, he withdrew in pique. In his place the Admiralty appointed two German naturalists, John Reinhold Forster and his son George. William Wales and William Bayly accompanied the expedition as astronomers. William Hodges was official artist.

Between them, these men provided illustrations and data to complement Cook's own meticulous account of what has been called the greatest single voyage in the history of the world. Enduring, according to the Forsters, "a series of hardships never before experienced by mortal man," Cook took his ships where no European vessel had gone before and kept them at sea for longer than had been thought possible. For one period Cook was at sea 122 days consecutively, for another 117 days, sailing once for ten thousand miles over strange seas without sighting land. No one before had ever explored the far southern waters of the Indian Ocean, or the Pacific, or the Atlantic. Cook was the first to cross the Antarctic circle, and indeed came very close

View of the Endeavour River with the *Endeavour* on shore being repaired. Now the site of Cooktown, Queensland.
Engraving by Will Byrne after Sydney Parkinson, 1770. From *An Account of the Voyages...for making Discoveries in the Southern Hemisphere*, ed. by John Hawkesworth, vol. III, 1773.

to discovering the Antarctic continent itself.

Cook proved beyond further doubt that a great fertile southern continent was an illusion, by sailing repeatedly over a great part of the area it was supposed to cover. In addition, he made two tremendous circuits of the southern Pacific. The first sweep took him from New Zealand to Pitcairn Island, to Tahiti, to the Friendly (Tongan) Islands and back to New Zealand. The second of these sweeps was the greatest exploratory voyage of its kind ever undertaken in the Pacific, extending thousands of miles from deep in the Antarctic almost to the equator and from New Zealand to east of Easter Island. In all, the three-year voyage covered some sixty thousand miles, two and a half times the circumference of the earth. It was a tremendous triumph for Cook, who was now acclaimed as one "who had done for geography and seamanship more in his voyages than any man since Columbus."

Cook's success, his ability to range over vast areas of ocean, to accurately chart new lands, to know where he had been and how to get back there, opened up new vistas for exploration that had been considered impossible. He owed some of his competence in navigation to recent advances in that science. In the 1760s it became possible to determine astronomically a ship's position at sea to within about half a degree of longitude, by the use of Hadley's reflecting quadrant or sextant, a pocket watch, and accurate astronomical tables such as those published by Nevil Maskelyne in the *Nautical Almanac*. The method, known as lunar distances, had for instance made it possible for Captain Wallis to chart the position of Tahiti accurately and for Cook to find it again.

Before a practical method of measuring longitude at sea was devised, a navigator made his landfall by "running down his latitude," which meant sailing east or west along the parallel of latitude of his destination until he sighted it. Longitude was measured by the precarious method of dead-reckoning: keeping a record of direction and speed and making allowance for current. With so much opportunity for miscalculation, islands were often missed and ships wrecked on shores that were not expected to be so close.

Now, for the first time, navigators could be equipped and trained to sail straight towards their intended landfall and to chart coastlines to within an accuracy of thirty nautical miles. Cook's First Voyage and the accuracy of his mapping demonstrated both the efficiency of this method and his mastery of it.

The disadvantage of the lunar distance method was its complexity. Whereas latitude, or north-south

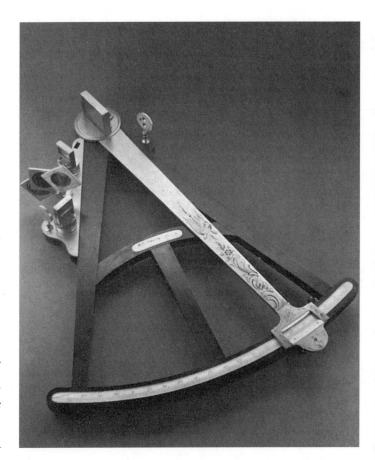

Hadley's octant or quadrant.
Made by F. Goater, London, c. 1780.

position, could be measured by quick reference to the sun at midday, calculation of longitude from the moon took much longer. A simpler method was necessary. In 1713 Isaac Newton had pointed out to a government committee that it should be possible to calculate longitude from a reliable clock giving

K1-Larcum Kendall's first chronometer. Construction: 1767-1769. Adjustments and Trials: 1769-1772.

the time at the Greenwich meridian, which when compared with local time, established by observation of the sun, would enable mariners to determine their east-west position. There was, however, no clock that would keep accurate time and be unaffected by storms at sea or by large changes of temperature.

It was over forty years before such an instrument was invented, by a self-taught Yorkshire clockmaker, John Harrison. It was a masterpiece of engineering and precision, with complex devices to compensate for temperature changes. In 1761, when the 72-pound clock was tested on a voyage to Jamaica, it lost just five seconds over the three-month journey, corresponding to a distance of under one mile. As Harrison proudly remarked: "There is neither any other mechanical or Mathematical thing in the world that is more beautiful or more curious in texture than this my watch."

On his second voyage Cook carried a copy of Harrison's fourth timepiece—a fraction of the size of his first—and, by careful comparison with his own astronomical calculations, established its accuracy in determining longitude to within about three miles. The problem of finding longitude at sea by a chronometer had finally been solved thanks to Harrison and to Larcum Kendall, who made Cook's replica (called K1) in 1769. Harrison's design eventually won for him in 1773 the greater part of the prize of twenty thousand pounds offered by Queen Anne in 1714 in "An Act for Providing a Public Reward for such Person or Persons as shall Discover the Longitude at Sea."

Cook mentions the watch in his journal many times. He described it as "our trusty guide the watch," writing that ". . . it would not be doing justice to Mr. Harrison and Mr. Kendall, if I did not own that we have received very great assistance from this useful and valuable time-piece." It enabled him to become the first commander of a ship in history to know almost precisely where he was for most of his time at sea.

The new possibilities for oceangoing vessels afforded by the chronometer would have been greatly lessened but for Cook's successful experiments in preserving the health of his men during their long spells at sea. One of his most remarkable achievements was the elimination of many of the sicknesses and diseases which commonly plagued ocean voyages. Above all he made great strides towards protecting his men from the scourge of the eighteenth-century seaman—scurvy. This ugly affliction, caused by lack of vitamin C, results in debility, depression, loss of teeth, haemorrhaging and death.

The mid-eighteenth century was a period when the Admiralty was being bombarded with suggested cures for scurvy, ranging from powders and pills to seawater and blood-letting. Cook's voyages presented an opportunity to experiment with some of the dietary and medical suggestions of the time. He was well supplied with a wide range of supposed antiscorbutics, including malt, sauerkraut (preserved cabbage), dehydrated soup and the sweetened, boiled juice of lemons and oranges.

Cook's voyages showed that scurvy was no longer to be feared if simple and effective precautions were strictly enforced. He was convinced that scurvy resulted from dirty conditions and bad diet. The ships, therefore, had to be kept scrubbed, ventilated and fumigated. The men had to wear clean, warm cloth-

K3-Larcum Kendall's third chronometer.
Construction: 1772–1774. Adjustments and trials: 1774–1776.

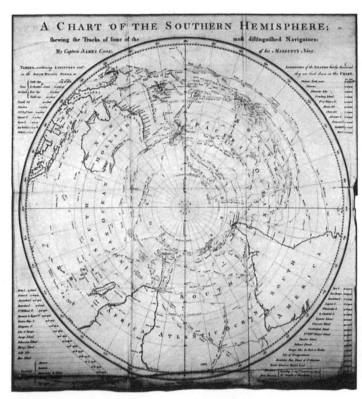

"A chart of the Southern Hemisphere" depicting Cook's and earlier explorers' tracks in the Southern Seas.
Engraving by William Whitchurch from *A Voyage towards the South Pole*... by James Cook, vol. 1, 1777.

Joseph Banks (1743–1820).
Engraving by J. R. Smith, 1773, after a portrait by Benjamin West.

ing. And every day, Cook forced down their throats unsalted soup, sauerkraut and fruit juice. More important still was his emphasis on fresh food and water. Whenever the ships touched on a suitable shore, the water was replenished and crewmen were sent foraging for fruit, vegetables, berries and green plants. They gathered wild celery, scurvy grass and wild cabbage, all of which, Cook reported, "were found to eat very well either in Soups or Sallids." They also brewed spruce beer.

The customary diet of seamen in the eighteenth century consisted of salt meat, worm-eaten oatmeal and rancid butter washed down with a gallon of beer per man — the daily allowance — or half a pint of rum or brandy in its place. Yet at first Cook's seamen rejected their captain's dietary innovations. They had an instinctive suspicion, as Cook wrote, of any food "out of the common way altho' it be ever so much for their good." They had to be cajoled, sometimes tricked and even bullied into accepting the new food. Although he had always been sparing in his use of the lash, Cook nevertheless had men flogged for refusing the prescribed meals. He preferred to use guile, however. From his own experience on the lower deck he well understood "the temper and dispossition of seamen." On the *Endeavour* he overcame the crew's resistance to sauerkraut by letting them know that it was being served exclusively to the officers, knowing that when sailors "see their Superiors set a value upon it, it becomes the finest stuff in the world." The strategy worked. Within a week the sauerkraut was so popular that it had to be rationed.

A result of Cook's meticulous attention to the welfare of his men was that on his second voyage to the Pacific and after remaining at sea for longer than anyone else in history, though there were outbreaks of scurvy only one man was lost through sickness. On his return Cook set down his methods in a paper to the Royal Society, of which he was elected a Fellow on 29 February 1776. It was a high honour in itself to be numbered among the leading scientists and scholars of the kingdom. The paper took the form of a letter to Sir John Pringle, President of the Royal Society and himself, as Surgeon-General of the Army, an authority on scurvy.

This contribution to the health of seamen earned Cook the Copley Gold Medal, the Society's award recognizing the summit of intellectual achievement. The occasion was the only time the medal has been granted for advances in the study of nutrition. It was his proudest triumph. "It is with real satisfaction," he wrote at the end of his journal of the Second Voyage, "and without claiming any merit but of attention to my duty, that I can conclude this account with an observation which facts enable me to make, that our having discovered the possibility of preserving health amongst a numerous ship's company for such a length of time in such varieties of climate and amidst such continued hardship and fatigue will make this voyage remarkable in the opinion of every benevilent person when the dispute about a southern continent shall have ceased to engage the attention and divide the judgement of philosphers."

Chapter Four: Planning the Third Voyage

Upon Cook's return from the Second Voyage he was appointed to a lucrative and undemanding position at Greenwich Hospital, a residence for disabled and retired Navy men. It seemed only fitting, after the rigours of the past six years, that he should be allowed to comfortably retire. The map of the Pacific, however, was not yet complete, and there was already talk of organizing another major voyage to explore its northern waters. Cook was the obvious choice as leader, but the Admiralty was reluctant to burden him further. As his first biographer recounts, "The benefits he had already conferred on science and navigation and the labours and dangers he had gone through were so many and so great that it was not deemed reasonable to ask him to engage in fresh perils." Cook solved their dilemma by volunteering. At a dinner given by the First Lord of the Admiralty to discuss the proposed expedition, Cook, it was reported, "was so fired with the contemplation and representation of the object that he started up and declared he himself would undertake the direction of the enterprise."

Cook was intrigued by the prospect of being the first to uncover one of the few remaining ocean mysteries — the geography of the region where the Pacific meets the Arctic. Here, it was hoped, he would discover the long-sought navigable waterway between the Pacific and the Atlantic oceans, the legendary Strait of Anian or Northwest Passage. For three centuries, since the discovery of America, European navigators had searched in vain for the location of

Map of Tartary, 1588.
By Abraham Ortelius showing the "Strait of Anian" and the northwest of the American continent.

this short route to the riches of the Orient.

The legend of the Strait of Anian was based on a misunderstanding by cartographers and geographers of Marco Polo's reference to Ania or Arian, thought to be a province in northeastern Asia. The waterway which was supposed to lead to it was called the Strait of Anian. On maps and in sailors' stories it appeared as an ice-free passage, easy to sail, stretching from the Atlantic to the Pacific, with its possible entrance at any considerable opening from the Gulf of California to Hudson's Bay. Many lives and fortunes had been spent in pursuit of this mystery.

Englishmen, in particular, had sought this passage

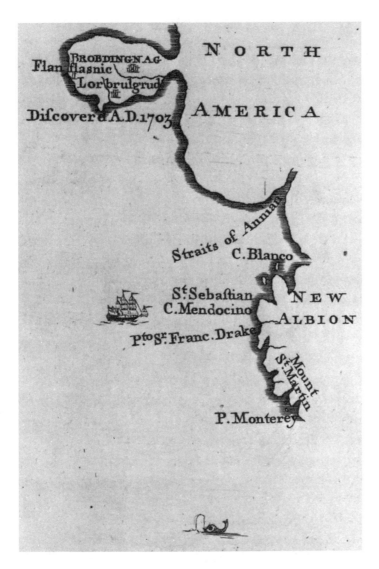

Brobdingnag — the country where Lemuel Gulliver found himself after a storm; sited in the approximate area of the province of British Columbia.

between the oceans, as an alternative to the southern routes to the Orient (through the Strait of Magellan and around the Capes) which had for centuries been controlled by England's rivals, Spain and Portugal. Elizabeth I's seamen and their successors, including such great explorers as Henry Hudson, John Davis, Martin Frobisher and William Baffin, had carried out the search from the hostile North Atlantic. In the process, through fog, shipwreck, storm and ice, they had discovered much of the northern and eastern coasts of Canada.

Yet still in Cook's day, English knowledge of the North Pacific coast of America remained vague and uncertain. No English ship had been there since Francis Drake had claimed New Albion for the Crown in 1579. Accordingly there was room for a great deal of conjecture and imagination; it was roughly in the area of present-day Oregon and British Columbia that Jonathan Swift placed the fantasy land of Brobdingnag in *Gulliver's Travels*.

Exploration of the Pacific coast of North America by Russian and Spanish navigators did little to dispel the idea that a waterway passed through or around the continent to the Atlantic. Reports of the discovery by Vitus Bering in 1728 of the strait later named after him, and of other Russian voyages along the coast of Alaska, renewed hopes that a passage did exist.

Rumours of Spanish discoveries also increased speculation. An inland passage allegedly travelled in 1592 by Juan de Fuca, a Greek pilot in the service of

Spain, appeared on eighteenth-century maps between 47° and 48° north latitude, close to the strait which bears his name today. Another route, with its entrance at 53° north (close to Skidegate Inlet between the major islands of the Queen Charlottes) was said to have been sighted in 1640 by a Bartholomew de Fonte. The two passages, whose discovery, it was suspected, Spain had tried to keep secret, sometimes appeared together on maps. They were shown running side by side, de Fuca's depicted as a long narrow channel of open sea and de Fonte's as a chain of rivers and lakes emptying into Baffin Bay. Such stories were believed to be true by many eighteenth-century geographers. Nonexistent inter-oceanic passages and other geographical fantasies, such as an inland sea covering much of today's British Columbia, were common features on eighteenth-century maps of the North Pacific.

The question of a northern passage had taken on renewed importance for England by the time of Cook's return from his second expedition. The East was becoming increasingly important to the future of the British Empire. India was becoming a springboard for the expansion of commerce through southeast Asia to China and the Pacific. In particular, the trade in Chinese tea was progressing rapidly. It was now essential, therefore, for Britain to establish the shortest possible route to the Far East, and a government reward of twenty thousand pounds awaited the first British ship to discover the passage. Both Commodore Byron and Captain Wallis had been sent on the search, but both had failed to reach even the northwest coast of America. Now, a man who was the undoubted master of the Pacific had appeared. The Admiralty was confident that if the fabled Strait of Anian existed, Cook would find it.

Cook's voyage in search of the Northwest Passage was to be part of a double-pronged attack. While his expedition was undertaking its search from the North Pacific, another probe, under the command of Richard Pickersgill, who sailed with Cook on his second voyage, was to be made once more from the Atlantic, through Baffin Bay. The objective of both expeditions was to reach the Arctic Ocean, where it was hoped they would meet. As eighteenth-century geographers were convinced that the ocean could not freeze, ice was not thought to be an obstacle to navigation in the Arctic. For this reason, neither expedition was equipped for work in the ice.

Cook's instructions for the search clearly indicated that this voyage would be the longest and toughest expedition he had yet undertaken. He was to take two small sailing-ships from the English Channel down the length of the Atlantic to the Cape of Good Hope. From here, he was to make for sub-Antarctic waters in search of island bases for future British voyages. Then for ten thousand miles he was to battle the winds as he crossed the bottom of the world to Tasmania and New Zealand. Next he was to sail to the Society Islands and Tahiti, and then, in the summer months, having taken on provisions and completed repairs, he was to explore the unknown waters of the North Pacific in search of the Northwest Passage and if possible to return through it to England.

There was no doubt as to Cook's choice of ship for this voyage across the whole breadth and almost the entire length of the Pacific. He would sail again in the *Resolution*, which, just two months after Cook's return, was already refitting for this second "voyage to remote parts." She was to be accompanied by another Whitby collier, the 298-ton *Discovery*, the smallest of all Cook's ships, under the thirty-three year-old Charles Clerke, who had been Second Lieutenant on the *Resolution* on the previous voyage.

Many officers from that voyage, in fact, had chosen to sail with Cook again. Some had been with him from the beginning. Clerke, indeed, had sailed as Master's Mate in the *Endeavour*, earning promotion during the voyage to Third Lieutenant. A farmer's son from Essex, keen for excitement, he was always cheerful, talkative and amusing, with an "open, generous Disposition." Clerke was a close friend of both Cook and Sir Joseph Banks. He had already crammed a wealth of experience into his nineteen years at sea. After entering the Navy in 1755 at the age of fourteen, he had seen action in the West Indies, where he miraculously survived being blown from the mizzen-top into the sea. Before beginning his service with Cook, he had sailed around the world on the *Dolphin* as midshipman under Byron.

The Master's Mate on that voyage, John Gore, who had also sailed with Wallis as well as Byron, was now First Lieutenant on the *Resolution* and at forty-

Map of the Pacific Northwest, pre-Cook (1752).
Thomas Jeffery's concept of de Fonte's alleged findings in northwest North America.

six the oldest officer in the company next to Cook. Gore, from Virginia, was one of the half-dozen Americans who sailed on the Third Voyage. He had joined the Navy in the same year as Cook and had sailed with him on the *Endeavour*, rising during the voyage from Third to Second Lieutenant. This was his fourth voyage around the world.

James King, aged twenty-six, Second Lieutenant on the *Resolution*, was sailing with Cook for the first time. The son of a Lancashire clergyman, he had entered the Navy at the age of twelve, earning promotion to Lieutenant in 1771. Like Cook, he had served under Palliser on the Newfoundland station. He had then studied science in Paris and at Oxford, where he so impressed the Professor of Astronomy that the professor recommended him for Cook's Third Voyage. His experience as a seaman, combined with his ability as an astronomer, made him of great service to Cook. As an officer and colleague he was well liked. One of the midshipmen wrote of him, "In short, as one of the best, he is one of the politest, genteelest and best-loved men in the world."

In contrast to King was the *Resolution's* rough spoken Master, William Bligh, also sailing with Cook for the first time and destined to become as famous as his captain. He had entered the Navy as an ordinary seaman and now, at the age of twenty-one, as Master responsible only to the captain, he had an excellent opportunity to make a name for himself. Cook made a good deal of use of his Master and evidently had confidence in Bligh's ability at reconaissance, charting, surveying, sounding, and reporting. His surveying work was so valuable that he received one eighth of the profits of the official account of the voyage. However, he was also vain and opinionated, with a hasty and violent temper that did not endear him to his colleagues or subordinates.

Perhaps the most openly disliked member of the company, however, was John Williamson, Third Lieutenant on the *Resolution*. He appears, from different accounts of the voyage, as a quick-tempered, self-righteous character with harsh opinions concerning the treatment of native people. He was to clash several times with Cook on this matter during the voyage. One of the midshipmen wrote of him,

Capt. Charles Clerke, R.N. (1743–79).
Portrait in oils, 1776, by Sir Nathaniel Dance-Holland.

Capt. John Gore, R.N. (c. 1730–90).
Portrait in oils, 1780, by John Webber.

"Williamson is a wretch, feared and hated by his inferiors, detested by his equals and despised by his superiors, a very devil, to whom none of our midshipmen have spoke for above a year: with whom I would not wish to be in favour, nor would receive an obligation from, was he Lord High Admiral of Great Britain." Able-seaman William Griffin called him "a very bad man and a great Tyrant." No other officer on any of Cook's voyages was so damned.

On the *Discovery*, Clerke's First Lieutenant was James Burney, aged twenty-one, a promising seaman who had sailed on the *Resolution* before, first as a seaman and then as midshipman and Second Lieutenant. He had returned from America to join Cook once more. His sister, the famous novelist Fanny Burney, wrote of his eagerness to join the voyage that "there is nothing Jem so earnestly desires as to be of the party: and my father (a friend of Lord Sandwich) has made great interest at the Admiralty to procure him that pleasure." A valuable officer, he was later to become an admiral.

Among the eleven midshipmen on the voyage, three were destined to achieve distinction. James Trevenen, a sixteen-year-old Cornish boy from the Royal Naval Academy at Portsmouth, had been recommended to Cook by Captain Wallis, and his appointment to the *Resolution* was his first seagoing experience. During the voyage he was to prove himself clever, generous, warmhearted and hardworking. He was killed at the age of thirty when fighting for the Russians against Sweden. Edward Riou, eighteen, on the *Discovery*, was also to die in battle. He was killed at Copenhagen in 1801 serving under Nelson, who wrote of him, "In poor, dear Riou the country

Capt. William Bligh, R.N. (1754–1817).
Engraving by J. Condi after a portrait by J. Russell.

Capt. James King, R.N. (1750–84).
Engraving by Francesco Bartolozzi after a portrait by John Webber.

has sustained an irreperable loss." He was the only one of Cook's men to be honoured by a monument in St. Paul's Cathedral. His fellow midshipman, George Vancouver, had sailed with Cook before on the Second Voyage as a thirteen-year-old able-seaman. In 1792 he began his own great voyage to the Pacific Northwest during which he made an accurate survey of the coast northwards from 30°, examined the Strait of Juan de Fuca and the Gulf of Georgia and circumnavigated Vancouver Island. His journal of that voyage with its constant references to Cook emphasizes the importance for these young men of a training under Cook and the admiration they had for him.

Affection and respect for Cook is evident as well in the journals and notes kept during the Third Voyage by members of the crew. Many of them had sailed with him before, including half a dozen who had been on both of his previous voyages. These veterans were: Robert Anderson, quartermaster on the *Endeavour* and now on his second voyage in the *Resolution* as gunner; William Harvey, Master's Mate and former midshipman; William Collett, who from able-seaman had now become Master-At-Arms and Cook's personal servant; John Ramsay, forty-four, back as an able-seaman after a spell as cook on the *Resolution;* William Peckover, gunner on the *Discovery,* who was later to serve with Bligh on the *Bounty* and be set adrift with him during the famous mutiny. Then there was Samuel Gibson, now a Sergeant, who as a marine on the First Voyage had been flogged by Cook for deserting the ship and taking to the hills with a Tahitian girl. He could speak native languages better than any man on board except the surgeon.

Some thirteen others had been with Cook on the epic Second Voyage. In the *Resolution*, these were: Henry Roberts, Master's Mate, who was to serve as principal assistant hydrographer to Cook. He drew the charts for engraving in the printed edition of the voyage, published in 1784; the American Bo'sun, William Ewin; the quick-tempered Irishman Patrick Whelan, former quartermaster in the *Resolution* and now serving as able-seaman; John Cave, able-seaman from Durham, who was to desert at Macao on the

Capt. Edward Riou, R.N. (1758?–1801).
Colour sketch.

voyage home; William Watman, able-seaman, aged forty-four, who had followed Cook out of a comfortable retirement at Greenwich Hospital. He was, wrote King, "an old man" who "had been 21 years a Marine . . . belov'd by his fellows for his good and benevolent disposition"; Richard Hergest, able-seaman, a twenty-two-year-old Londoner; Alexander Dewar, the ship's Scottish clerk. In the *Discovery*, the veterans were: William Lanyon, Master's Mate, from Cornwall; Richard Collett, twenty-three, able-seaman on the Second Voyage and now Master-at-Arms; the carpenter Peter Reynolds; the cook Robert Goulding, formerly carpenter's mate; and Edward Barrett, twenty, from London, formerly cook's mate.

There were some interesting newcomers as well. John Ledyard, an adventurous American aged twenty-five, had sailed in 1773 to Europe, where he had tried, unsuccessfully, to enlist in a British regiment at Gibraltar. He had now, somehow, become Corporal of Marines on the *Resolution*. After the voyage he refused to serve against his countrymen during the American War of Independence and deserted in 1782 while serving on the North American station. In 1784 he returned to London and set out to walk across Siberia, down to Nootka Sound and then to Virginia. He actually reached Irkutsk before he was arrested in February 1788 and returned to the Polish border. He died at Cairo on an expedition looking for the source of the Niger.

Another American, Nathaniel Portlock, sailing on the *Discovery* as Master's Mate, and George Dixon, the armourer, were both later to receive their own commands in an expedition to develop the Pacific Northwest fur trade during which, in 1787, Dixon discovered the Queen Charlotte Islands. An ordinary seaman in the *Discovery*, Joseph Billings, later be-

Capt. Nathaniel Portlock, R.N. (1748–1817).
Painting in oils.

came a captain in the Russian Navy and explored the extreme northeastern parts of Asia and the Bering Sea. Another adventurer was Heinrich Zimmerman, seaman on the *Discovery*, a twenty-five-year-old German belt maker who had begun to wander in 1770

John Webber, R.A. (1751-93).
Portrait miniature in oils by J. Daniel Moffet.

and had worked at various trades in Geneva, Lyons, Paris and London. He wrote of himself that "the natural courage of a native of the Palatinate" determined him "to adopt a seafaring life" and volunteer into the *Resolution*. Another volunteer was George Gilbert, eighteen, sailing as an able-seaman in place of his father, Joseph Gilbert, former Master of the *Resolution*. Then there was Benjamin Lyon, a former watchmaker from London, whose old skills were put to use during the voyage in an attempt to repair the *Resolution's* chronometer. A surprising talent was revealed by James Clevely, the *Resolution's* carpenter, who made detailed drawings of scenes on the voyage from which, on his return home, his brother John worked up a series of paintings.

Others who attract attention were the newly commissioned Irish Lieutenant of Marines, Molesworth Philips, twenty-one, who twice challenged Williamson to a duel. Bligh described him as "a person who never was of any real service the whole voyage, or did anything but eat and sleep." He had little training himself and no ability to train his men, but there was little for him or his marines to do except guard duty.

The surgeons too had little call on their services during the voyage, thanks to Cook's close supervision of the health of his men. They showed talent in other directions, however. William Anderson, surgeon on the *Resolution,* had been the ship's Surgeon's Mate on the Second Voyage. Described by King as "by far the most accurate and inquisitive person on board," he was an enthusiastic self-taught naturalist, interested in all branches of natural history and in native languages as well. His ability to interpret new words and his ethnological observations were of great assistance to Cook. He went to great pains to draw up vocabularies for the different peoples visited on the voyage; the one on Tahitian speech was later printed.

Sadly, he was to suffer increasingly throughout the voyage from tuberculosis.

Anderson's First Mate was David Samwell, a parson's son from Wales, whose interests were more romantic than scientific. He had a definite eye for the ladies, writing two weeks before the *Resolution* sailed, "For my part I live as happy as I cou'd wish only that one's cut off from the Society of the Dear Girls." His colleague on the *Discovery*, William Ellis, "a genteel young fellow" who had been educated at Cambridge and St. Bartholomew's Hospital, proved to be a talented water-colourist.

As with previous voyages, an official illustrator was commissioned by the Admiralty, so that in Cook's words, "we might go out with every help that could serve to make the result of our voyage entertaining to the generality of our readers as well as instructive to the sailor and scholar." The artist John Webber, the son of a Swiss sculptor and educated in art at Berne and Paris, was to make this the most profusely illustrated of all the voyages. Another official appointment was William Bayly, astronomer, who had sailed in the *Adventure* on the Second Voyage. He was later to become headmaster of the Royal Academy at Portsmouth. With him on the *Discovery* was David Nelson from Kew Gardens, sent by Banks to collect plants and seeds. Clerke described him in a letter to Banks as "one of the quietest fellows in nature." He was also to sail in the *Bounty* in 1787 to supervise the gathering of breadfruit in Tahiti, and, like Peckover, was set adrift with Bligh. He died of

Omai.
Drawing in graphite with sanguine wash by Sir Nathaniel Dance-Holland, 1774.

Death of Capt. James Cook at Kealakekua Bay, Hawaii on the morning of 14 February, 1779.
Engraving: figures by Francesco Bartolozzi, landscape by W. Byrne after an oil painting by John Webber.

fever and exposure in Timor, 20 June 1789.

And then there was Omai, a Society Islander, brought to England at his own request by Captain Furneaux in the *Adventure*. Sir Joseph Banks had dressed him as a fashionable English gentleman and introduced him to English society. During his two years in London, he had attended the opera, been the guest of nobility and gentry, and dined in company with Samuel Johnson. He had taken the fancy of Fanny Burney when he visited her brother James, the

Discovery's new First Lieutenant who could talk to him in his native language. He had also learned a certain amount of English during his visit, enough to upset court decorum by greeting King George III with a cheery "How do King Tosh." Cook was now to take him home again with many presents, including port wine, gunpowder, muskets and bullets, a suit of armour, a hand-organ, toy soldiers and a globe of the world. He could look forward to impressing his people with his gifts and stories of life in England.

For Omai's shipmates, the voyage promised disparate things: for the officers and midshipmen, honour and promotion; for the crew, hardship and discomfort. Yet even on the lower deck there were many who had joined just to sail with Cook. The voyage had the appeal of adventure, giving each of them a sense of their special importance. As Samwell was to write, "It is an article of faith with every one of us that there never was such a collection of fine lads take us all in all got together as there was in the *Resolution* and *Discovery*."

Part II: The Third Pacific Voyage

Chapter Five: Plymouth to Palmerston

By early summer 1776, preparations for the Third Voyage were almost complete. King George III granted Cook an hour's audience to wish him Godspeed, and the expedition was given a grand official farewell. Cook noted in his journal for June 8:

The Earl of Sandwich, Sr Hugh Palliser and Others of the Board of Admiralty paid us the last mark of the extraordinary attention they had all along paid to this equipment by coming on board to see that everything was completed to their desire and to the satisfaction of all who were to embark on the voyage. They and several other Noblemen and Gentlemen honoured me with their company at dinner and were saluted with 17 guns and three cheers at their coming on board and also on going a shore.

In July both ships were ready at Plymouth, seeming small and insignificant amidst the great fleet of warships, troop-carriers and supply vessels bound for the war in the American colonies. The Declaration of Independence was signed on 4 July, just eight days before the *Resolution* sailed.

Although Britain and the United States were to be at war during the entire voyage, Cook's expedition was considered to be so important for the advancement of knowledge that his ships were exempted from capture or attack. Benjamin Franklin, American Minister to the court of France, realized that American privateers might seize and hold the two vessels as hostages on their voyage home and wrote the following letter requesting free passage for Cook:

To all Captains and Commanders of armed Ships acting by Commission from the Congress of the United States of America now in War with Great Britain. —

Gentlemen,
A Ship having been fitted out from England before the commencement of this War to make Discoveries of new Countries in unknown Seas, under the conduct of that most celebrated Navigator Capt. Cook; an undertaking truly laudable in itself, as the Increase of Geographical Knowledge facilitates the Communication between distant Nations, in the exchange of useful Products and Manufactures and the extension of Arts, whereby the common enjoyments of human life are multiplied and augmented, and Science of other kinds increased to the benefit of mankind in general. This is therefore most earnestly to recommend to everyone of you that in Case the said Ship, which is now expected to be soon in the European seas on her return, should happen to fall into your Hands, you would not consider her as an enemy, nor suffer any Plunder to be made of the effects contained in her, nor obstruct her immediate return to England by detaining her or sending her into any other part of Europe or to America, but that you would treat the said Captain Cook and his people with all civility and

kindness, affording them as common friends to mankind all the Assistance in your power which they may happen to stand in need of. In so doing you will not only gratify the generosity of your own dispositions, but there is no doubt of your obtaining the approbation of the Congress and your other American Owners.

I have the honor to be,
Gentlemen,
Your most obedient
humble Servant
B. Franklin

Given at Passy,	*Minister Plenipotentiary*
Near Paris, this	*from the Congress of the*
10 Day of March	*United States to the Court*
1779	*of France*

The *Resolution* had already been at sea for two weeks before Captain Clerke was able to follow in the *Discovery*. The unfortunate Clerke had been imprisoned for debt after agreeing to act as guarantor for the payment of the debts of his brother, a captain in the Royal Navy who had sailed to the East Indies. Burney had brought the ship down to Plymouth while Clerke struggled to obtain his release. He finally joined the *Discovery* on 1 August through the intercession of Banks, to whom the "castaway but everlasting, grateful obliged" Clerke wrote an exuberant note of thanks: "Huzza, my boys heave away—away we go—adieu my best friend; I won't pretend to tell you how much I am your grateful and devoted C. Clerke." Lamentably he was never to be free from the effects of his days in the King's Bench Prison, for there he contracted tuberculosis, which was gradually to weaken him throughout the voyage. For the present, however, it was a jubilant new Captain Clerke who set out to rendezvous with Cook at Cape Town.

Meanwhile, the journey south had revealed serious defects in the *Resolution*. As Cook was to learn throughout the voyage, the Royal dockyards, then notoriously inefficient and corrupt, had skimped on fitting her out. In the high winds and heavy seas of the Atlantic she now began to leak badly. All the quarters were wet and some of the stores were spoiled. The mizzenmast was found to be so seriously cracked that it could not bear sail.

However, there were some light moments. On 1 September the *Resolution* crossed the equator, and Cook, who had lost no time in drilling the crew in his methods for ensuring good health and hygiene, condoned the usual revels and antics associated with crossing the line. The surgeon, Anderson, was emphatic in his disapproval, writing in his journal of the "old ridiculous ceremony of ducking those who had not crossed the Equator before. This is one of those absurd customs...which every sensible person... ought to suppress instead of encouraging."

The *Resolution* anchored at Cape Town on 18 October 1776 and the *Discovery* arrived there three weeks later after a rough passage during which a marine had been lost overboard. This was Cook's fourth visit to the colony and he was entertained with his officers by the Governor. He even had his portrait painted by Webber, who had already completed a full-length study of him before they left London. Anderson passed his time by exploring the countryside for plants and insects. He was greatly impressed by the ability of the Dutch settlers to "raise such plenty in a spot where I believe no other European nation would have attempted to settle." Meanwhile, the ships were overhauled as thoroughly as possible and enough provisions for two years were taken on board. The Harrison chronometer was checked for accuracy and found to be performing excellently. The crew dined on fresh mutton, greens and soft bread and enjoyed the last comforts of civilization before returning, as Clerke wrote, "to the old trade of exploring." One of the crew, William Hunt, armourer, was returned to England for passing bad coinage.

At the Cape, additions were made to the considerable number of livestock brought from England. The *Resolution,* which carried all of the animals because the *Discovery* had no room, was turned into a floating farmyard and now housed four horses, three bulls, four cows, two calves, fifteen goats, thirty sheep, a peacock and innumerable pigs, hens, turkeys, rabbits, geese and ducks. Omai gave up his cabin to the horses, "with raptures," as Cook wrote to Banks and Sandwich, commenting as well that "nothing is wanting but a few females of our own species to make the Resolution a compleate ark." The animals were to be distributed among the Pacific islands as a gift from "Farmer" George III. Many died, but the pigs survived, multiplying to such an extent that their descendants, known as "Captain Cookers," still roam New Zealand.

Both ships left the Cape on 30 November 1776, already a month behind schedule. They sailed south into the high winds and huge seas of the Roaring Forties, bound for desolate Kerguelen Island in the South Indian Ocean via the Prince Edward Islands, discovered by the French explorer Marc-Joseph Marion du Fresne as recently as 1772. Before they reached Kerguelen, the fog was so thick that the ships had to fire their guns intermittently to keep in touch. The care of the livestock gave great trouble, and after a week, sheep and goats began to die.

On his way out on the Second Voyage, Cook had learned at the Cape that Yves de Kerguelen had reported sighting land in latitude 48° south and in the longitude of Mauritius (he was some twenty degrees of longitude in error). Cook had searched for it on his first Antarctic sweep and missed it by about ten degrees. He now ran down the latitude of Kerguelen Island until he sighted it on 24 December 1776, in latitude 48½° south, longitude 68° 40' east. The following day they anchored in Christmas Harbour. Cook wrote:

> I immidiately despatched Mr. Bligh the Master in a boat to Sound the Harbour, who on his return reported it to be safe and commodious . . .

and great plenty of fresh Water Seals, Penguins and other birds on the shore but not a stick of wood.

The island was so bare of vegetation that Cook called it "The Island of Desolation." On its barren hills grew neither wood nor shrubs, and fewer than twenty kinds of plants were found, although Anderson discovered another weapon against scurvy in a new kind of cabbage; this he named "Pringlea antiscorbutica" after Sir John Pringle, then President of the Royal Society. The sandy shores, however, abounded with penguins, fur seals, ducks and sea birds (Plate 1). The penguins were docile and easily clubbed; their flesh provided fresh meat and their oil was used for lamps. Later, thanks to the accuracy with which Cook charted his position, the seals would be exterminated by sealers.

On 30 December 1776, after six days of exploring Kerguelen's rocky coastline for a good harbour for future British voyages, Cook set sail across the bottom of the world towards Tasmania and New Zealand. On the way, the *Resolution* lost her topmast in a squall and another had to be fitted while the ship rolled violently as she ran before the gale.

At Tasmania, which they reached on 26 January 1777, they found wood for heating and cooking and the grass which they urgently needed for the cattle. Five marines who had saved their liquor ration got drunk on shore duty during the few days' stay and were severely punished. Edgar noted that they "made themselves so beastly Drunk, that they were put motionless in the Boat and when brought on board oblig'd to be hoisted into the Ship."

Abel Tasman had discovered Van Dieman's Land as early as 1642, without encountering any of the inhabitants. The next European to visit the island was Marc-Joseph Marion de Fresne who, in 1772, did meet some Tasmanians but was driven from the island when a misunderstanding arose, after which he was prevented from landing again. After cruising the coast for six days, Marion departed for New Zealand, and Van Dieman's Land was not visited again until Capt. Tobias Furneaux's equally brief stay in 1773. Furneaux saw many fires in the woods but met none of the natives.

Cook expected only a brief stay, like his predecessors'. The morning after anchoring he sent some men ashore along with a marine guard to collect supplies. The marines were along as a precautionary measure, for Cook had also observed smoke issuing from the woods and, not knowing anything about the inhabitants, was unwilling to take risks. However, none of the Tasmanians made an appearance. By evening the provisions had been brought on board and Cook prepared to sail the following day. A windless morning foiled this plan, but the day brought some interesting experiences.

Omai was demonstrating his superior skills at fishing and the woodcutters were collecting more spars for the ships when a marine, who was cooking some food a little distance from his fellows, was suddenly confronted by a small group of men. According to

PLATE 1

A View of Christmas Harbour, in Kerguelen's Land [Kerguelen Island].

Samwell, "The man [was] struck with Terror & Astonishment at this unexpected Appearance [and] ran towards the Boat crying out 'here they are here they are!'" The others, instantly on their guard, advanced in a group towards the strangers who had materialized from the woods. For many of the crew, these were the first flesh-and-blood "Indians" they had ever met, and undoubtedly they were fearful, for apart from their superstitions they were aware of their proximity to the cannibals of New Zealand.

It must have been a sheepish contingent of Englishmen who found themselves advancing towards ten naked, unarmed Tasmanians who were approaching them "without the least mark of fear and with the greatest confidence immaginable." Within moments the tension relaxed and each scrutinized the other. "Upon the whole these indians have little of that fierce or wild appearance common to many people in their situation," observed Anderson, "but on the contrary seem mild and chearfull without reserve or jealousy of strangers." When, the following day, their visitors included some equally naked women and children who displayed the same gentle dispositions, the general consensus was that here was a people who "have few, or no wants, & seemed perfectly Happy, if one might judge from their behaviour, for they frequently would burst out, into the most immoderate fits of Laughter & when one Laughed every one followed his Example Emediately."

On the other hand, the English noted with surprise that the Tasmanians showed far less curiosity about their visitors than was to be expected from a people who had never before seen Europeans. The instinctive reaction of astonishment and terror displayed by the marine was precisely the type of response that the eighteenth-century discoverer was used to eliciting from native peoples. The Polynesians, the "model primitives," were inquisitive about everything new and would blithely steal anything in sight. But the Tasmanians, like the Patagonians of Terra del Fuego whom Cook met on his First Voyage, were not easy to understand. Did they exemplify the Natural Man that eighteenth-century writers admired—unaffected and uninterested in material possessions? They certainly had little, not even ornaments, just stone tools; and they went naked. Or was their lack of curiosity or desire to possess English trinkets a sign of dullness of mind? The eighteenth-century English mind vacillated between admiration for the Noble Savage whose virtue was in being ingenuous, and disparagement for the slothful savage for displaying no ingenuity. Clerke decided that the Tasmanians were harmless, cheerful, simple people, but Anderson saw their primitiveness as a lack of progress and pronounced them deficient in "personal activity," indifferent, inattentive and even less inventive than the "half animated inhabitants of Terra del Fuego." Cook observed both their cheerfulness and indifference, but refrained from drawing conclusions.

The recording of such differing perspectives of the same experiences makes the journals of Cook's Third Voyage a prism-mirror for eighteenth-century attitudes towards the "uncivilized nations." In their journals Cook, King, Anderson, Clerke and Samwell all react in their own way, from the analytic approach of Anderson to the empiric deductions of Cook. Each was stimulated by different things at different times. When viewed together, the journals display a spectrum of beliefs, attitudes and facts that reflect European cultural values just as much as they present a picture of the new cultures they are writing about.

Webber's engravings also reveal his individual perspective, as well as the conventions of eighteenth-century views of man and theories of art. Plate 3, "A Woman of Van Dieman's Land," is as complex and provocative a statement as that of any of the journalists. It may be viewed by an ethnographer as a visual record of a prototypical Tasmanian woman, for it depicts her features, the style of her haircut and how she carries her baby, but the portrait also projects an attitude which is Webber's, although it reflects some of the words of the journalists. Webber has constructed the entire picture around the woman's gesture of holding the hand of her peacefully sleeping baby. The significance of this pose is embedded in the eighteenth-century European view of the savage state where the natives were characterized by Jean-Jacques Rousseau and his followers as "Noble Savages" living in a southern Arcady and modelled after the early Greeks; others, especially Thomas Hobbs, saw them as "Ignoble, Depraved Savages" whose life was mean, nasty, short and brutish. Depraved Savages were believed to be destitute of any human feeling. They killed and ate their enemies and demonstrated no affection for their kinsmen, especially their children. But the Tasmanian women that the English met were observed "to carry their babies in a manner that show'd some degree of tenderness." This observation set the affective tone of Webber's portrait: he elevates the woman to the stature of a Noble Savage. In keeping with neoclassical conventions, Webber maintains the unity of mood and expression by not inserting any contradictory details. In this portrait he has not permitted human imperfections to intrude upon an epic concept. His neoclassical vision has not been corrupted by, nor has he conveyed any of, the critical attitudes of some of his companions.

The profuse recording of opinions, observations and drawings was based on only a few hours of contact with the Tasmanians. This style of reportage, though typical of the "grand tour," became, in the hands of some competent observers, the beginnings of anthropology. All the journalists made extensive notes on native manufactures, especially weapons. Anderson collected comprehensive word lists everywhere he went, and Cook detailed the natives' political organizations. As geographers and hydrographers these men were unquestionably the most skillful map-makers of their century. As explorers they also made valiant attempts to map the ideas — religious, moral, political, economic — of the peoples they met. If their cultural explorations were not as skillful and as complete as their mapping of coastlines, nevertheless the journals from Cook's three voyages constitute primary source material for the study of the peoples of the Pacific. In the case of the Tasmanians, they are one of the few surviving early records of an extinct civilization. Tasmania became a penal colony in 1803, and in 1876 the last aboriginal Tasmanian died.

Cook left Tasmania on 30 January 1777 for New Zealand, where he intended to complete the stocking of the ships' wood and water in preparation for the voyage to Tahiti. There was a second reason for this destination: a grim one. During the course of his exploration of Antarctica on his Second Voyage, Cook had stopped at New Zealand's Queen Charlotte Sound and at nearby Dusky Sound four times. At one point, Cook and his sister ship, the *Adventure*, had been separated and Captain Furneaux had put into Dusky Sound. There, at Grass Cove, a small party of sailors from Furneaux's ship fell victim to Maori cannibals. Furneaux did not investigate the massacre and so, as the ships were never reunited, Cook only learned of the event in Cape Town, en route home. "I shall make no reflections on this Melancholy Affair," he had written in his journal at that time, "untill I hear more about it. I must however observe in favour of the New Zealanders that I have allways found them of a Brave, Noble, Open and benevolent disposition, but they are a people that will never put up with an insult if they have an opportunity to resent it." Cook was sailing to Queen Charlotte Sound and the Maori

PLATE 2

A Man of Van Diemen's Land [Tasmania].

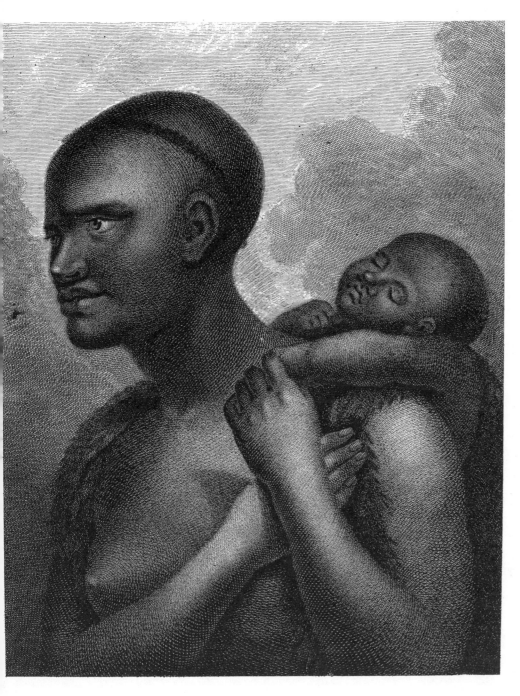

PLATE 3

A Woman of Van Diemen's Land [Tasmania].

people, whom he had come to know very well, to find out what had happened.

The ships put into Queen Charlotte Sound on 11 February 1777, and Cook soon discerned apprehension among the few Maoris who ventured out to the ships. Their long acquaintance with Cook was barely acknowledged as they waited to see what action he would take. They fully expected Cook to exact *utu* (retribution) in accordance with their concept of justice: an eye for an eye.

Cook assured them that his intentions were friendly, and discreetly postponed his enquiries for five days. He then journeyed to nearby Grass Cove in Dusky Sound, where the gruesome incident involving Furneaux's men had taken place. With Omai's help, he discovered what had happened:

> I met with my old friend Pedro who was almost continually with me the last time I was in this Sound.... [He] told us that while [Furneaux's] people were at victuals with several of the natives about them some of the latter stole or snatched from them some bread, & fish for which they were beat this being resented a quarrel insued, in which two of the natives were shot dead, by the only two Muskets that were fired, for before they had time to discharge a third or load those that were fired they were all seized and knocked up the head. They pointed to the place of the Sun when this happened, and according to it it must have been late in the afternoon: they also

PLATE 4

An Opossum of Van Diemen's Land [Tasmania]

shewed us the spot where the boats crew sat at Victuals, and the place where the boat laid which was about two hundred yards from them with Captain Furneaux's black servant in her.

Cook had already decided against punitive measures, even though when this became evident one man named Kahura fearlessly boasted of his participation in the massacre. Cook's reticence led to some criticism from his own crew, who were vexed by the insolence of a self-confessed murderer wandering freely aboard ship. Even Omai importuned Cook to act:

Why do you not kill him. You tell me if a man kills another in England he is hanged for it. This man has killed ten and yet you will not kill him. tho a great many of his countrymen desire it and it would be very good.

Cook remained firm in his decision not to play judge and executioner. He even came to admire Kahura's pluck. But he did note that "if ever they made a Second attempt of that kind, they might rest assured of feeling the weight of my resentment."

Only one of Webber's New Zealand drawings was engraved for publication with the journals. It is a rather nondescript view of an island fortress that the officers visited one afternoon during their two-week stay (Plate 5). Many portraits and views from New Zealand had been published with the accounts of the first two voyages, which may explain why there was only one published for the Third Voyage. It is equally plausible that Webber was discouraged from making many drawings by the now unhappy associations which New Zealand called to mind.

The stay in New Zealand was of necessity a brief one. Uppermost in Cook's thoughts were the orders from the Admiralty that instructed him to arrive in 65° north latitude by June. Already it was the end of February and he was several weeks behind schedule. He must reach Tahiti without delay, then set course for the coast of North America. However, an insufficient knowledge of the shifting trade winds between New Zealand and Tahiti foiled his plan, for although he had made this passage before, it had been in a different season. This time the winds failed him utterly. Languid, tedious days—more than thirty of them—found the cattle nearing starvation and the crew growing increasingly disgruntled.

At the end of March they sighted land. Cook had happened upon a new group of islands which were later to be named after him. The most southerly of these "Cook Islands," Mangaia, afforded them no safe anchorage, so they could not get grass for the cattle or fresh food for the crew. The islanders thronged the shore and some swam out or came in canoes. However, only one man, named Mourua,

PLATE 5

The Inside of a Hippah [fortress], in New Zeeland.

PLATE 6

A Man of Mangea [Mangaia, Cook Islands].

was bold enough to board, and Webber drew him wearing, as an ear ornament, a knife which Cook had presented to him (Plate 6). They had little more success at the small island of Atiu, or at Manuae, which they reached on 6 April. It was here Cook made his decision:

Being thus disapointed at all these islands, and the summer in the northern Hemisphere already too far advanced for me to think of doing any thing there this year, It was therefore absolutely necessary to persue such methods as weas most likely to preserve the Cattle . . . and save the ships stores. . . . I therefore determined to bear away for the Friendly Isl^{ds} where I was sure of being supplied with everything I wanted.

It was a much happier group of men who set out for the Friendly or Tongan Islands. They had before them the prospect of several months of languid days, but this time in the company of the women of Polynesia. A brief stop at uninhabited Palmerston Island whetted their appetites for the fuller pleasures of the Friendly Islands, for here there were coconuts to be gathered, fish to be caught in abundance, and all kinds of birds. "Men of War and Tropic Birds, Boobies, Noddies and Egg Birds, . . . stood to be stroak'd about the Bows of the Trees, a certain and indisputable proof of their perfect Ignorance of every thing resembling a Human Form," wrote Clerke. Unfortunately, they were to lose their innocence—and their lives—to the voracious appetites of English seamen.

Chapter Six: Tonga or the Friendly Islands

Nomuka, the first stop in the Friendly Islands, was a South Sea paradise. David Samwell, the Surgeon's Mate, was caught in its spell and wrote that Nomuka "may be said to realize the poetical descriptions of the Elysian fields in ancient writers, it is certainly as beautiful a spot as Immagination can paint." Here were beautiful ponds and lagoons where the sailors could hunt or just walk along paths shaded from a sultry sun by bowering branches. The country was everywhere "exceedingly well cultivated & [had] the appearance of a beautiful Garden," wrote Samwell. Nature gave abundantly, and man was enabled to live "in the State most agreeable to his Nature undisturbed by those Passions, those Vultures of the Mind, that are found to distract & torment him in artificial Society."

Cook sent the emaciated cattle and the observatory tents ashore accompanied by a guard of marines because during his previous visit in 1774, when the Nomukans had laid eyes on Europeans for the first time, not only had they accosted his men and relieved them of their muskets and tools but also an old lady had severely admonished Cook for demanding the return of the stolen goods. This same woman had offered Cook her daughter as a concubine, and his refusal also earned him a tongue-lashing. "What sort of man are you thus to refuse the embraces of so fine a young Woman," she admonished. The young woman's beauty he could withstand, Cook had confessed, "but the abuse of the old Woman I could not and therefore hastened into the Boat."

On arrival at Nomuka this time, Captain Cook was greeted with generous gifts from the chiefs (*eiki*) of the island. Immediately a marketplace was set up where coconuts, pigs, breadfruit, yams and plantains were traded for glass beads, cloth and items made of iron (Plate 7). People from nearby islands arrived in canoes laden with food for trade, and aboard ship a sumptuous feast soon obliterated the memory of the many lean weeks without fresh food.

But not everything changed hands by the orderly means of the marketplace. As Williamson tells us, "The natives soon gave us a specimen of their happy genius in the art of pilfering." The Tongans made away with whatever took their fancy, which was just about everything from turkeys to clothing to Captain Clerke's cats. Cook and Clerke were hard pressed to invent equally ingenious methods of punishment and dissuasion. Clerke finally discovered that flogging the culprit, shaving one half of his head and unceremoniously depositing him overboard was as good a deterrant as any, but still the thefts persisted.

The diarists, though, acquitted the Friendly Islanders of any taint of vice which their thieving implied, and transformed the thievery into a positive virtue, for the English thought that the Tongans were unlike their suspiciously "Hungry Acquaintances," the New Zealanders, or the harmless, unsophisticated Tasmanians, and were, to use Anderson's words, "in every respect almost as perfectly civiliz'd as it is possible for mankind to be. They seem to have been long at their ultimum...." Consequently, any moral

PLATE 7

A View at Anamooka [Nomuka, Tonga].

qualms the English might have had were to be suppressed to fit this considered opinion. Thievery came to be seen as an index of the Tongans' lively curiosity rather than as the mark of the indolence and avarice of a savage mind.

The diarists' high opinion of the Tongans flowed from the unfailing cordiality and hospitality of their hosts, who not only provided a fine cuisine but also entertained them day and night with sports and musical performances, which everyone agreed "would have met with universal applause on a European Theatre." All around them they observed strong, active people who were always mild and friendly; who seemed entirely free from all the "base passions" such as greed, envy and lust, and who were exceptionally graceful and well proportioned. In fact, there were "hundreds of truly Europaean faces & many genuine Roman noses amongst them." In their persons and their houses they were decent and clean, and everywhere there was evidence of an industrious people. They were cultivators, not hunters, and they concentrated on the civilized arts of poetry, dance and music with an aesthetic intensity that matched the intelligent skills which they used to bring the natural world under their control. Their qualitites "do honour to the human mind," acclaimed King, "[and] prove abundantly that these people are far remov'd from a savage state."

Only one caviller opposed these eulogies and this, surprisingly, was Samwell, the parson's son whose love of the "nubile nymphs" so belied his heritage. All of the diarists witnessed some "Scenes of Barbarity," but most glossed over them for reasons which will shortly become clear; Samwell, the exception, recorded that the innocent common people were entirely at the mercy of powerful chiefs who often showed a wanton disregard for their subjects' property and persons. Samwell did not doubt that the common people were virtuous and deserved the praise his fellow officers lavished on them, but "barbarous" was his word for any society which exalted a man "to such an unnatural Pitch of greatness" that laws ceased to govern his behaviour towards his subjects.

Cook was not blind to the impunity with which some chiefs wielded power, but he tended to look away and to focus instead on trying to understand the Tongan system of government. On all three voyages, Cook carried with him the Earl of Morton's *Hints offered to the consideration of Captain Cooke* ... on what to make note of when encountering new nations. Religion, morals, order, government, distinctions of power, police and tokens for commerce were prominent subjects for inquiry. Cook's interest in order and government was as much personal as it was professional, and his curiosity to understand the distribution of power in Tonga emerged as a keen preoccupation. "It was my interest as well as inclination to pay my Court to ... great men," Cook declared, and in Tonga he was not disappointed in finding more than one pretender to the title of "King of all the Friendly Isles." And since he perceived the

Friendly Islands to be fundamentally a "peaceful nation," free both from wars between the islands and petty, personal disputes between families, Cook took it upon himself to spend the leisurely weeks in Tonga gathering as much information about Tongan society as he could. In fact both Cook and Anderson, with Omai's assistance as translator, produced an anthropological and scientific account of Tongan society that was equal in scope and perceptivity to the close observations of Tahiti that Joseph Banks made on Cook's First Voyage.

The first person to introduce himself as "King of all the Friendly Isles" was a man named Finau, who arrived when the ships were anchored at Nomuka. According to Burney:

> Finow, a tall handsome man, appeared to be about 25 years of age; had much fire and vivacity with a degree of wildness in his countenance that well tallied with our idea of an Indian Warrior. add to this he was one of the most active men I have ever seen.

Cook and Finau entertained one another with gifts and feasts that befitted their high positions. Within two weeks, however, the island's food supply was exhausted, and Finau invited Cook to come to Ha'apai, a group of islands to the northeast.

Finau guided the *Resolution* and the *Discovery* around the treacherous coral reefs which form a submerged linkage between the islands of the Ha'apai group. They anchored off the lovely island of Lifuka and were "immediately surrounded by a multitude of Canoes." Finau took full charge, escorted Cook before a crowd gathered on shore and delivered a harangue about which Cook remarked:

> The purport of this speach as I learnt from Omai, was that all the people both young and old were to look upon me as a friend who was come to remain with them a few days, and that they were not to steal or molest me in any thing, that they were to bring hogs, fowls, fruit & c^a to the ships where they would receive in exchange such and such things.

Here, indeed, was a man of consequence. People listened to him in respectful silence, and Omai was convinced of his regal status. Finau's reassurances and open, friendly manner also convinced Cook of the appropriateness of the name "The Friendly Isles" which he had conferred on the islands on his last voyage. The Ha'apaians were indeed friendly, and the days at Lifuka were given over to the pleasures of music, sport and dancing.

Finau's hospitalities began the very next day. Early in the morning he again conducted Cook before a large crowd gathered in an open space.

> I had not sit long before near a hundred people came laden with Yams, Bread fruit, Plantain

Cocoanuts and Sugar Cane which were laid in two heaps or piles on our left.... Soon came a number of others from the right, laden with the same kind of articles which were laid in two piles on our right; to these were tied two pigs and six fowls, and to those on the left six pigs and two turtle.... As soon as every thing was laid in order and to shew to the most advantage, those who had brought in the things joined the Multitude who formed a large circle around the whole.

The crowd began to cheer and sing as wrestling, stick combat and boxing contestants entered the ring (Plate 8). The English admired the Tongan physique and were impressed with the wrestlers (Plate 9) for their "prodigious exertion of strength" and their musculature "of such a size as would serve for an artist to draw from a living Hercules," which is precisely what Webber did. But the most amazing feature of the sports was the ordered and civilized manner in which they were conducted. The contests were entered into with such good humour that the victor did not crow nor the vanquished suffer shame.

The Ha'apaians showed such skill and agility in martial arts that some of Cook's men decided to enter the contests, but they were hopelessly defeated by the Tongans at wrestling, and in the boxing matches not one escaped being knocked down.

The seamen not only took a thorough drubbing but also were shocked when into the circle formed by hundreds of cheering and laughing Tongans stepped "a couple of lusty wenches who without the least ceremony fell to boxing, and with as much art as the men." But English protests fell on deaf Tongan ears, so one gentlemen, "perhaps smitten got up and interfered which produced Loud Shouts of Applause as well as much mirth among the Spectators."

When the sports were over, Finau formally presented Cook with one of the piles of food brought during the early morning procession. The other pile he gave to Omai. This spectacular gift filled four boats and "far exceeded any present I had ever before received from an Indian Prince." Cook could not hope to make a sufficient return and, indeed, when he attempted to do so, Finau only plied him with further gifts. His prestige, both in the eyes of Cook and the Ha'apaians, was much enhanced by these displays so graciously received by the English guests.

Two days later, on 20 May, the entire day was taken up with grand entertainments. Cook, at Finau's request, put the marines through their paces. Carrying a British flag on a long staff, they marched before an assembled crowd of more than two thousand. Finau was particularly keen to see what muskets were capable of and seemed satisfied at their effectiveness when a canoe was pierced clear through. The Tongans then proceeded to perform their now famous paddle dance, the precision and beauty of which left the marines looking like gawky schoolboys. The dance was accompanied by music and singing in perfect harmony. Cook tells us that the performance "so far

PLATE 8

The Reception of Captain Cook, in Hapaee [Tonga].

PLATE 9

A Boxing Match, in Hapaee [Tonga].

exceeded anything we had done to amuse them that they seemed to pique themselves in the superiority they had over us."

The English determined to assert the superiority of their civilization by staging a display of fireworks, for which an enormous crowd had assembled by evening. Samwell wrote:

> They were all waiting with eager Expectation when about 8 o'Clock the first Piece, a Water Rocket, was played off; it is not possible to convey by words an adequate Idea of the Astonishment & Surprize they expressed at seeing fire burn in the Sea, & now & then even diving under water, then rising & shining with redoubled Lustre & at last going off with a sudden & unexpected Explosion.... when the Baloons went off with a horrid Explosion close to their Ears it was hard to say whether their Terror or Surprize was greater.... their applause and admiration were expressed by a continual Roar during the whole Exhibition... In short every thing appeared to them like Enchantment.

English magic had its intended effect.

The evening concluded with a *po me'e* or evening of dancing. Tongan dancing utterly enchanted the English. The journalists wrote long complicated descriptions of it, and Webber tried to capture its aesthetic and sensual charm (Plates 10, 11). But behind this particular *po me'e* lay a treacherous conspiracy.

Over the previous few days, the high chiefs of Ha'apai had been hatching a plan to surprise Cook and his men, kill them and take the ships' booty. Finau joined the conspiracy and decreed that the best time would be during the morning when Cook's men were preoccupied with the entertainments. Some of the lesser chiefs, however, felt that the cover of night would be essential. Finau was inflamed by this questioning of his authority and cancelled the plan. His vanity saved Cook his life, although Cook never knew it. The story was recorded by William Mariner, an Englishman whose life was spared after a successful ambush of his ship, *Port au Prince,* in this same harbour of Lifuka in 1806.

But at the *po me'e* that night the plotting and quarrelling were artfully concealed. The evening was illumined by burning palm leaves and scented with fragrant oils burning like candles on coconut leaves. "The greatest number of beautiful girls that we had yet seen" thronged the beach, wrote Samwell. Bamboo sticks pounded a slow, rhythmic beat and soft voices lifted in a plaintive melody.

Wearing only a soft tapa-cloth skirt, the women danced first, their long, delicate fingers providing much of the expressive movement of the dance (Plate 11). The men danced after them, displaying more vigorous, though still agile and graceful, movements (Plate 10). Singing and recitation accompanied both performances, and Anderson expressed much regret in not being able to describe the language "which would doubtless afford much information as to the genius and customs of these people." There was no doubt in his mind that the dance was designed to elevate the nobler sentiments. Again, he stressed the cultivated virtues of these Noble Savages.

A week of lazy, balmy days on Lifuka was suddenly enlivened with the arrival at the *Resolution* of a very formidable character. Like Finau, he claimed to be "King of all the Friendly Isles." We had reason to doubt this," said Cook, "but [the people] stood to it and now for the first time told us that Feenough [Finau] was not the King." Omai was alone in insisting that Finau was the greater of the two chiefs.

By his behaviour and the respect accorded him by the people, Cook gradually deduced that this man, who was called Fatafehi Paulaho, was indeed the higher ranking chief. This judgement was confirmed when Finau, conspicuously absent upon the arrival of Paulaho, made deep obeisance before this chief and avoided eating in his presence. Finau was observing tabu-respect, the customary submissiveness of a man of lower rank. Paulaho was, in fact, the Tu'i Tonga, the highest ranking man in all of Tonga. He was said to be descended from the great creator god, Tangaloa, who fished the Tongan Islands from the sea and fathered the first Tu'i Tonga.

Omai was crushed. For three weeks he had cultivated Finau's friendship with gifts from his precious cache brought from England, in the end only to lose face. As it turned out, Finau was nevertheless a man of some consequence, being of the Tu'i Kanokupolu, the chiefly lineage which exercised secular power

PLATE 10

A Night Dance by Men, in Hapaee [Tonga].

PLATE 11

A Night Dance by Women, in Hapaee [Tonga].

PLATE 12

Poulaho [Paulaho], King of the Friendly Islands.

whereas the Tu'i Tonga's lineage possessed religious power. Cook referred to Finau as Paulaho's "generalissimo."

In contrast to Finau's leanness and dynamism, Paulaho was the "most corperate plump fellow we had met with." (Webber's engraving of Paulaho, Plate 12, shows none of his mass.) The journalists preferred Finau's noble, manly physique to the shapeless bulk of the "Lord of All," whom they portrayed as hedonistic and indolent; but Paulaho nonplussed them by asking some very shrewd questions. According to King,

> he now ask'd our business there, where we came from, & where we were going, questions which had never been put to us before; to satisfy him a map was produced & explained to him, & by his explaining it again to his followers, he shew'd that he was capable of receiving & comprehending subjects which we considered above their faculties in their present state.

Paulaho now took full command of the English visitors. He invited Cook to accompany him to the great island of Tongatapu, the seat of his power.

On 10 June 1777 the *Resolution* and the *Discovery* anchored in Nuku'alofa Harbour, Tongatapu. Here began a peaceful, month-long holiday made luxurious by the gracious attentions of King Paulaho and his kinsmen. Aristocratic ceremony marked the daily encounters between the English officers and the

Tongan royal family. Captain Cook was "quite charmed with the decorum that was observed," confessing that "I had no where seen the like, no not even amongst more civilized nations." However, some ceremonies, such as that of the Kava Ring, were not to his liking (Plate 13). It was not so much that Cook did not enjoy kava, for the brew created a pleasant languor, but the manner of preparing it quite quenched whatever thirst he had. Because the hard, gnarly kava root had to be softened, the King's attendants first chewed it before immersing it in water to squeeze out the juice. Cook managed to drink an occasional ceremonial cup, but not with as much relish as Paulaho drank the wine and brandy Cook gave him when he dined aboard the *Resolution*.

Evenings on Tongatapu were devoted to singing and dancing. Once Cook visited the sacred royal burial ground (Plate 14), but he usually occupied his days with walking into the countryside and watching the people, noting what they cultivated, what they manufactured, and the methods they used. His observations were detailed and interesting.

Perhaps Cook's most valuable contribution to what later anthropologists might call "Tongan ethnography" was his account of the Inasi Ceremony (Plate 15). "Inasi" means "share," and this particular Inasi witnessed by Cook seems to have been in honour of King Paulaho's son, who was to be given a "share" of his father's authority.

The Inasi was considered so sacred that it was hedged with tabus. Only high ranking men and women could participate, and Cook was only allowed to watch it from behind a bamboo curtain. Not a man to be commanded, Cook kept stealing away to get a better view. People stopped him, pressing him to return to his appointed spot, which he did at first, not being sure of the consequences of breaking tabu. At the moment when Paulaho and his son were to eat together, to share food and symbolically share the father's power, all were instructed to sit with their backs turned and eyes downcast "as demure as Maids." "Neither this commandment nor the remembrence of Lot's wife discouraged me from facing about," Cook confessed, but it was to no avail, for too many people stood in his way.

As the ceremonies resumed, Cook was determined to see them in full. He wrote:

I resolved to peep no longer from behind the Curtin but to make one of the number in the Ceremony if possible; with this View I stole out of the Plantation and walked towards the Morai; I was several times applied to go back by people who were passing to and fro, but I paid no regard to them and they suffered me to pass on. When I got to the Morai, I found a number of men seated on one side.... When I got into the midst of the first Company I was desired to sit down which I accordingly did.... I was several times desired to go away, and at last when they found that I would not stir, they, after some seeming consultation, desired I would bare my

PLATE 13

Poulaho [Paulaho], King of the Friendly Islands, Drinking Kava.

PLATE 14

A Flatooka, or Morai [sacred burial ground], in Tongataboo [Tongatapu].

shoulders as they were, with this I complied, after which they were no longer uneasy at my presence.

Naked to the waist and hair streaming, Captain Cook participated in the Inasi. However, in doing so he violated proprieties set down by his own civilization. His fellow officers were appalled; as Williamson commented:

> We who were on ye outside were not a little surprised at seeing Captn Cook in ye procession of the Chiefs, w^t his hair hanging loose & his body naked down to ye waist, or with his hair tyed; I do not pretend to dispute the propriety of Captn Cook's conduct, but I cannot help thinking he rather let himself down.

The long ceremony finally drew to a close and Cook made preparations to leave Tongatapu. He was tempted to stay longer to observe a funeral ceremony, but it was to last five days and the tides and winds were favourable and beckoning him to continue his planned journey to Tahiti. His inquiries into Tongan affairs of state had been remarkably thorough. Although his genealogical investigation uncovered some of the complex relations between people who held power, he was unable to fathom the subtler Tongan distinctions between political and religious ranking which accounted both for Finau's authority to take Paulaho's life if he should prove despotic and for the fact that Paulaho's sister, the Tamaha, and her children all ranked higher than Paulaho himself. Cook's conclusion on the entire matter was that Paulaho was a benevolent, wise ruler and that the friendly and basically happy disposition of his people proved that his government was not despotic.

Cook made Paulaho a present of horses, cows and goats in the expectation that this sensible man would care for them and their progeny. On 10 July 1777 the ships at last set sail for Tahiti.

But one last stop was to be made among the Friendly Islands, this one at Eua where Cook believed he could collect fresh water. Just as Samwell fell in love with Nomuka, Edgar was caught in the spell of Eua:

> This little Island may be stiled the Montpelier of the Friendly Isles—the Land is higher than any of the others, and affords some of the most Romantick & beautiful Valleys in the World.

Even Cook was charmed by its pastoral beauty. He climbed the island's highest hill to look over "medows ... adorned with rufts of trees and here and there plantations." The view inspired an unusually proprietary thought, and he wrote, "I could not help flatering myself with the idea that some future Navigator may from the very same station behold these Medows stocked with Cattle, the English have planted on this island."

PLATE 15

The Natche [Inasi], a Ceremony in Honour of the King's Son, in Tongataboo [Tongatapu].

PLATE 16

A Woman of Eaoo [Eua].

John Webber must have been similarly entranced. His portrait of "A Woman of Eaoo" (Plate 16) displays a romantic vision equal to the eighteenth-century Italian neoclassical masters'. The woman's gentle beauty stands as a symbol of their peaceful sojourn in the Friendly Islands.

Chapter Seven: Tahiti and the Society Islands

By the time of Captain Cook's Third Voyage, Tahiti had already captured the romantic imagination of Europe. Or perhaps it is more accurate to say that the eighteenth-century European imagination had captured Tahiti in its romanticizing. The French explorer Louis de Bougainville, writing of Tahiti in 1768, said that in Tahiti "one would think himself in the Elysian fields." The place was so idyllic, confided Bougainville, that "I thought I was transported into the Garden of Eden... We found companies of men and women sitting under the shade of their fruit trees [and] everywhere we found hospitality, ease, innocent joy, and every appearance of happiness amongst them." Joseph Banks, travelling with Cook on his First Voyage, wrote in 1773 that the men and women of Tahiti were so exquisitely formed that they might "defy the imitation of the chizzel of a Phidias...," while their country "was the truest picture of an Arcadia *of which we were going to be kings that the imagination can form*" (author's italics). Such romantic language is an example of metaphorical thinking, a type of imaginative reasoning which J. C. Beaglehole has described as "preposterous sublimity" or "nonsense on stilts." In these quotes from Bougainville and Banks, the Christian vision of Paradise before the Fall has been fused with an eighteenth-century cultural classicism that harkened back to the ancient Greeks for a model of the simple, pure life. This romantic attitude was a reaction to the Industrial Revolution in England, a longing to return to the simple, virtuous country life. Although it did not go unchallenged, this idyll of life on a South Sea island amongst Noble Savages was a compelling eighteenth-century fantasy with religious overtones. Its themes were beauty, uncorrupted human nature and childlike innocence.

Though not a man given to idealizing the Noble Savage, Captain Cook was, nevertheless, a man of his time. Echoing the panegyrics of Bougainville and Banks, Cook had written in his First Voyage journal that "in the article of food these people may almost be said to be exempt from the curse of our fore fathers; scarcely can it be said that they earn their bread with the sweat of their brow, benevolent nature has not only supply'd them with necessarys but with abundance of superfluities." When he was leaving the Society Islands in 1775 during his Second Voyage, Cook expressed deep regret that he would never again see these beautiful islands or meet his many Tahitian friends. He had become attached to a young Tahitian named Odiddy and wanted to take him back to England, but in 1775 there was no prospect of his making a third voyage, so with no hope of returning Odiddy to his native land, he had decided against the idea. To Cook, Odiddy was "a youth of good parts, and like most of his countrymen, of a Docile, Gentle and humane disposition."

The Third Voyage journal accounts of Tahiti disappoint the expectant reader who has enjoyed Banks's *Endeavour* journal and Forster's Second Voyage Observations. The journalists present mere adumbrations of this fabled land, all giving the same reason for not providing extensive comments. To quote King: "I shall not pretend to give any particular description of what is in this Island, as it has been so often visit'd all that has already been done & by those who have had leisure to attend to these things." So there is not much critical comment during this Tahitian interlude and, with a few noteworthy exceptions, the journalists wrote about it as though it were a well known land.

Cook's visit to Tahiti in 1777 had the specific purpose of repatriating Omai. That he should be performing this task at King George III's behest was ironic indeed, for Cook's opinion of Omai was in sharp contrast to his affection for Odiddy. Omai "is ...a downright blackguard," Cook wrote in 1775. But upper-class English society did not agree, and Omai had been taken in hand by Joseph Banks, pampered and lionized. After he left London, a pantomime was written about him that brought audiences flocking to see it, and a Royal Command performance had been given.

Cook's opinion of Omai was modified, but not significantly altered, by this fame. Omai sometimes amused Cook, though more often he proved annoying. He adopted haughty airs, was foolish with his wealth, had no political astuteness, and exhibited the flamboyance of a peacock. In his journal, Cook constantly monitored Omai's behaviour, but a touch of Cook's dry humour often lightened his most exasperated moments with Omai, as in the following scene:

Omai and I prepared to pay [the] young cheif a formal Vesit. On this occasion, Omai, assisted by some of his friends, dressed himself not in English dress, nor in Otaheiti, nor in Tongatabu nor in the dress of any country upon earth, but in a strange medly of all he was possess'd of.

Omai received a great deal of attention in the journals. In contrast to the journalists' lack of interest in writing about Tahitian society, there developed an interest in individual characters. Omai shared the spotlight with an "Unseen Presence" at Vaotepeha Bay, a beautiful Raiatean "princess" named Poetua and a very well defined personality at Matavai Bay, a chief called Tu.

An atmosphere of holiday anticipation spread through the *Resolution* and the *Discovery* during the early morning of 12 August 1777 as they came in sight of "Otaheiti." Samwell remarked:

Omai sat all day on the forecastle viewing his Native Shores with Tears in his Eyes. Our two New Zealand Friends were not backward in testifying their Joy at seeing the Promised Land of which they had heard so much from Omai & others in the Ship.

Cook made his first stop at Vaitepeha Bay, intending to obtain basic supplies there before settling in at Matavai Bay, his preferred anchorage. There was a great stir of excitement on that first day when the

Tahitians discovered that the English had red feathers to trade. Red feathers from the blue-crowned lory (*Vina australis*) were prestige possessions, being even more highly prized than iron. They were essential to all religious rites (Plate 17) and were associated symbolically with the high chiefs and the gods. The red-plumed lory was not native to Tahiti, and the English had wisely stocked up on the feathers in Tonga, where they were not so highly valued. (Even the Tongans had to get the feathers from Fiji.) Cook observed on that first day that "not more feathers than might be got from a Tom tit would purchase a hog of 40 or 50 pound weight." However, as red feathers glutted the market, prices plummeted and by the end of the day, their value had decreased to a fifth.

Within hours of their arrival at Vaitepeha Bay, Tahiti, Cook learned that a number of important changes had occurred there. The young high chief he had met in 1775 had died and his ten-year-old brother was now the most powerful chief. But what most captured his attention was a report that two Spanish ships from Lima, Peru, had come to Vaitepeha since his last visit and that four Spanish missionaries had lived there for approximately fifteen months before returning to Lima. From the Tahitians, Cook gathered that the Spanish had accomplished two things: they had convinced the Tahitians of their superiority over the English, and they had gained not only their friendship but also their veneration. The missionary house, complete with iron nails, furnishings and discarded clothing, stood unmolested like a reliquary. The missionaries had maligned the English in a manner "absolutely unbecoming the dignity of any nation," wrote Anderson. Clerke, overflowing with sarcasm, denounced them for "their deportment [which] was quite in the Spanish Style, brim full of pomp and State." Clerke imagined a puffed-up "Signior" responding to a Tahitian's mention of England: "'Oh! I recollect the place you mean; there was a damn'd little, dirty, piratical State call'd England, but ... our omnipotent King ... destroy'd the Country, & ras'd the rascally breed from off the Face of the Earth. ....'"

The status of the English was, however, quite easily restored with gifts of tufts of red feathers to important chiefs (the Spanish did not have feathers) and by a display of fireworks, another exclusively English feat. To emphasize their position further, Cook performed a small act of desecration. In front of the missionary house stood a cross which the Spanish had engraved with their king's name and the date, 1774. To this Cook added his own inscription: "Georgius tertius Rex Annis 1767, 69, 73, 74 & 77."

Heavy rain during the first few days at Vaitepeha did not stop Omai from squandering his wealth on rapacious relatives, but it prevented Cook from going to investigate a Roman Catholic chapel which some of his men reported having seen from a distance. When he did go, Cook discovered that it was a *fata tupapou,* a "ghosthouse" in which was kept the embalmed body of the young chief who had died while the Spanish missionaries had been at Vaitepeha (Plate

18). Describing the *fata tupapau*, Cook said it was "uncommonly neat . . . covered and hung round with different Coloured cloth and Mats so as to have a pretty effect; there was one piece of scarlet broad Cloth of 4 or 5 Yards in length which had been given them by the Spaniards. This cloth and a few Tassels of feathers which our gentlemen took for silk, made them believe it was a Chappel, for whatever else was wanting their imagination supplied."

With a pressing invitation from King Tu to visit him in Matavai Bay, Cook took the ships to his old anchorage there, arriving on 24 August 1777. Cook had known Tu from earlier voyages and had not been impressed by him, but Tu's "Talents as a King" seemed to improve over the years and, after the Second Voyage, Cook was moved to judge him as "a Man of good parts, [who] had indeed some judicious, sensible men about him who I believe have a great share in the Government."

Although Cook called him "King" Tu, he was only one of three high chiefs in the Matavai Bay area in Tahiti. He was, in fact, a rather timorous character, given to importuning the English for military assistance against his enemies of the neighbouring island of Eimeo, or Moorea as it is called today. Ironically, because of the strategic position of Matavai Bay as the best anchorage in Tahiti for European trade vessels, Tu, with his astute solicitousness, gained a monopoly over the trade market. This gave him the power to defeat all his enemies and to declare himself King of Tahiti. This he accomplished in 1790, just thirteen years after Cook's last visit. Calling himself Pomare I, Tu consolidated his reign in 1801 by negotiating a pork trade with Australia, plump Tahitian hogs being exchanged for sturdy English muskets.

Within forty-eight hours of Cook's arrival at Matavai, his old friends had come with gifts. Omai, for a change, was "dress'd in his very best suit of clothes and conducted himself with a great deal of respect and Modesty" when Cook introduced him to King Tu "and the whole royal family." Cook had prepared a fine gift for Tu: "I gave him a Suit of fine linnen a gold laced hat, some tools and what was of more value than all the others, a large piece of red feathers and one of the Friendly islands bonnets" (Plate 12).

Odiddy, Cook's "Constant Companion" on the Second Voyage, also came, but Cook does no more than mention the fact of his visit in his journal. A puzzling response, until Samwell gives a clue:

Oididee the young fellow who had accompanied Captn Cook last Voyage in his Expedition to the s^o w and New Zealand, came to see us immediately on our arrival here & was much pleased with meeting with Captn Cook & some of his old Shipmates again. We had been told by those who had been in the Resolution last Voyage that he was a fine sensible young fellow, much superior to Omai in every respect, which made it some disappointment to us to find him one of the most stupid Fellows on the Island, with a clumsy awkward Person and a remarkable heavy look; . . . he frequently came on board to see us along with his Wife & was almost constantly drunk with Kava.

Here was a man whose qualities were known all over Europe because of the widely read translations of Cook's journals of the Second Voyage. Bayly also characterized Odiddy as "the most silly fillow" and King "the most stupid foolish Youth I ever say." Cook's silence was perhaps the most severe judgement.

In the ensuing few days, Tu's character also came under close scrutiny. King felt "the Conduct of the King & all his family was . . . very disgusting from their meanness in begging red feathers all around the Ship." Bayly was sure that a man he almost nabbed sneaking about in the observatory tent was Tu, because the would-be thief had hold of the small box Bayly kept his red feathers in and had shown only to Tu. Samwell reported that Tu, fearing reprisal from Captain Cook for a theft committed by one of his followers, "hid himself among the Bushes & would not be seen by him [Cook], till . . . Cook . . . went to him and told him he had nothing to fear." Cook noted that Tu stationed some of his men near the ships, where each morning they collected tribute from the women as they left their generous paramours.

Cook was drawn into Tu's politics almost immediately. A long-standing feud with the neighbouring island of Moorea suddenly flared up again shortly after the English arrived, and Tu's people wanted war. Tu himself vacillated, and Cook refused to assist, but the general feeling ran high in favour of retaliation. The necessary preparations were begun. The most important of these was the offering of a human sacrifice to the war god, Oro. Tu's powerful general, Towha, produced a suitable victim, an accused man from the lowest class of people.

Here was Cook's chance to witness and provide an authentic account of "this extraordinary and barbarous custom," for, as Anderson tells us, "the circumstances of these people sometimes offering human sacrifices is mentioned by Mr Bougainville from the Authority of the native he carried from the Island, but as these are cases often not credited unless the relater has ocular proof this was thought a proper opportunity to confirm it." And there was nothing more calculated to horrify (but at the same time titillate) the civilized mind than the stories of New World human sacrifices that had circulated since the sixteenth century. Cook's account of the sacrifice was as detailed and meticulous as his description of the Inasi ceremony at Tongatapu, and Webber's drawing (Plate 17) became one of the classics of the age. Although it depicts a terrible act, "Webber's Savages" stand beside Cook tall, virile and possessing all the attributes of noble, natural man. The landscape setting diminishes the horror of the battered corpse lying tied to a pole in the centre of the engraving. A row of grinning skulls, which could have lent a macabre feeling, instead merges into the background, muted behind a shaft of light that rests dramatically on Cook's face. In fact, the whole movement of the

PLATE 17

A Human Sacrifice, in a Morai, in Otaheite [Tahiti].

engraving is towards Cook.

The two-day ceremony of the human sacrifice took place on 1 and 2 September 1777 at King Tu's *marae*, a sacred area reserved as both a burial ground and a place for such ceremonies. Cook, Anderson and Webber silently watched the long rituals. It was a "horrid practise... detrimental to that right of self-preservation which everyone must be suppos'd to posess at his birth," wrote Anderson, though "the Custom would perhaps be less reprehensible upon the whole did it serve to impress any awe for the Divinity or reverence for religion upon the multitude." Anderson concluded that such sacrifices persisted because of the "grossest ignorance and superstition."

And when Tu's general, the old warrior Towha, asked Cook how he liked the ceremony, Cook did not mince words:

> As soon as [the ceremony] was over we made no scruple in giving our sentiments very freely upon it and, of course, condemned it. I told the Chief that this Sacrifice was so far from pleasing the Eatua [god] as they intended that he would be angry with them for it and that they would not succeed. Omai was our spokesman and entered into our arguments with so much Spirit that he put the Chief out of all manner of patience, especially when he was told that if he a Chief in England had put a Man to death as he had done this, he would be hanged for it; on this he balled out "Maena maeno" (Vile vile) and would not hear another word; so that we left him with as great a contempt of our customs as we could possibly have of theirs.

The English argued the merits of impartial justice, forgetting for the moment that even in England at that time it was rarely granted to the common man. To the Tahitian, justice was always partial, and it was inconceivable that a chief would ever be condemned to death. But considering that such a ceremony should have been unthinkable in this Garden of Eden, it is remarkable how little outrage it generated. Samwell, who found fault with the Tongan paradise, made no entries in his journal from 29 August to 20 September 1777. He resumed his commentary with the statement, "Since the beginning of this month no remarkable Transactions have occurred...." His earnest attentions were devoted to "the Girls."

The war party did not leave immediately after the sacrifice, for Tu still was not for war. Ten days later another human was sacrificed, though this time Cook was not present at the ceremony. Finally, on 17 September, the war fleet, led by Tu's general, Towha, set out for Moorea. Although Towha repeatedly requested reinforcements, Tu sent none. Within five days, with neither side gaining a clear victory, Towha had to negotiate an unsatisfactory peace treaty.

Towha returned in a rage. "The old Admiral was irritated to a Degree of Madness, he Abused [Tu] Every where," reported one of the *Discovery*'s crew.

The alliance between Towha, a passionate man of action, and Tu, an ambitious man of political cunning, broke down five years later and destroyed one of Cook's plans. Captain Bligh learned of the event in 1788 and recorded it in the log of the *Bounty*:

> [Tu] said that...the Imeo [Moorea] People joined with Tettowah, (the noted old Admiral called by Captain Cook Towah) and made a descent at Oparre [near Matavai Bay] that after some resistance by which many men were killed, he and all his People fled to the Mountains. The People of Imeo and those...under Tettowah now being masters of all their property, destroyed everything they could get hold of, among which were the Cattle, Sheep, Ducks, Geese, Turkeys and Peacocks left by Captain Cook in 1777. The Cows had eight Calves, The Ewes had ten Young ones.... Thus all our fond hopes, that the trouble Captain Cook had taken to introduce so many valuable things among them, would by me have been found to be productive of every good, are entirely blasted.

While Tu and Towha quarrelled, Cook's company amused themselves in much the same way as they had in Tonga. The days were leisurely and relations between the sexes were easy. They made excursions into the countryside, gorged themselves on Tahitian food, and were entertained day and night.

One evening Odiddy provided Cook with a rich feast of pork, fish and Cook's favourite pudding, *poe*, which he tells us was made from ground-up fruit, nuts, and taro mixed with coconut milk.

> We had but just dined when Otoo [Tu] came and asked me if my belly was full and on my answering in the affirmative, said "than come along with me." I accordingly went with him to his Fathers where they were dressing two girls in a prodigious quantity of fine cloth in a manner rather curious....

This cloth, together with a quantity of food, was presented to Cook as a gift from Tu's father. Drawing a well composed portrait that included this grotesque costume (Plate 19) must have presented Webber with a difficult aesthetic problem. He (or perhaps Bartolozzi, the engraver) solved it by posing the woman's arms in such a way as to draw attention to her face with its self-conscious, almost demure expression: an antipodean Gioconda.

Another evening was given over to an *arioi* dance-drama performed by Tu's sisters (Plates 20, 21). The *arioi* were a religious sect whose social function included both impromptu and formal ceremonial entertaining. They were skilled actors, pantomimists and dancers. For Cook, who thought that "the Misteries of most Religions are very dark and not easily understud even by those who profess them," the *arioi* were utterly baffling. He was shocked by the "licentious" nature of some of their performances, and the

PLATE 18

The Body of Tee, a Chief, as Preserved after Death in Otaheiti [Tahiti].

nobler sentiments of art were nowhere present as the *arioi* dancers deliberately attempted "to raise in the spectators the most libidinous desires." All of the journalists agreed that Tongan dancing was superior.

Cook characterized the sojourn on the island of Tahiti as one of uninterrupted cordiality. In contrast to Tonga, there was little theft, thanks to Tu. Even Omai prospered, in spite of his swindling relatives, and took possession of a fine Tahitian sailing canoe, complete with a crew and decorated with streamers, in which he planned to make a grand entrance at Huahine. Tu presented Cook with a beautifully carved canoe, a gift for the *arii rahi no Pretane,* or the Great Chief of Britain. Cook had to refuse the canoe because of its size, but he was touched by the gesture.

As he had never visited Moorea, Cook decided to stop there briefly to see for himself the scene of the abortive skirmish between Tu's people and Mahine, the high chief of that island. The visit had an unpleasant conclusion.

The harbour of Paopao proved to be one of the best in the whole of the South Pacific, and the setting was a monumental one of stupendous volcanic peaks. Beneath them lay the ruins of the recent war; "the trees were stripped of their fruit and all the houses were either pull'd or burnt down," observed Cook.

Cook was irritated from the start. The ships were infested with rats (Captain Clerke's cats had never been returned by the Tongans) and efforts to entice the rats to leave the ships via a bridge rigged up between ship and shore were largely unsuccessful.

A few days later trouble erupted. One of the goats intended for settlement on another island was stolen while it was grazing on shore, and Cook demanded its return. The next day another was taken. Later that night the first goat was returned, but the second one was not. The following day Cook threatened Chief Mahine and also sent out a search party, without success. Frustrated, he wrote: "I was now very sorry I had proceeded so far, and I could not retreat with any tolerable credet, and without giving incouragement to the people of other islands we had yet to visit to rob us with impunity."

He set out with a party of thirty-five men to recover the goat, but no one would admit to knowing its whereabouts. Cook then took an extreme and unprecedented action: he ordered his men to burn the Moorean's houses and to wreck their large war canoes. The next morning, the goat still not having been returned, he sent the carpenters to wreck several large canoes lying in the harbour, and to bring the planks on board to take to Hauhine for Omai's house. He did the same in the next harbour. At sunset, the goat was returned, and the following day the ships left for Huahine.

Gilbert summed up the general feeling about this display of destructive force which has puzzled historians to this day. "I can't well account," said Gilbert, "for Capt. Cook's proceedings on this occasion; as they were so very different from his conduct in like cases in his former voyages."

At Huahine, Omai came into his own. His arrival in the "Royal George Canoe" was eclipsed only by the appearance of the *Resolution* and the *Discovery*. The usual display of exuberant and light-fingered curiosity was diminished somewhat by the Tahitian passengers' exaggerated reports of Cook's rampage on Moorea.

Omai began acting responsibly, and earned Cook's commendation:

> We got ready to pay a formal visit to the young chief, Omai dress'd himself very properly on the occasion and prepared a very handsome present for the chief and another for his Eatua [god]., indeed after he got clear of the gang that surrounded him at Otaheiti [Tahiti] he behaved with much prudence as to gain respect.

The chiefs of Huahine designated a piece of land for Omai, and the ships' carpenters built for him a neat house, 24 feet by 18 feet by 10 feet high, "built of boards and with as few nails as possible, that there might be no inducement to pull it down" (Plate 22).

While this house was being constructed, Cook turned his mind to some irritating problems. The stock of bread was full of vermin and the ship was infested with cockroaches; as with the rats, all remedies were proving ineffective. In the midst of these worries a different species of pest made his way into Bayly's observatory and carried off a sextant. Cook described the man as a "hardened Scoundrel"

PLATE 19

A Young Woman of Otaheite [Tahiti], Bringing a Present.

PLATE 20

A Dance in Otaheite [Tahiti].

and, upon apprehending him, "punished him with greater severity than I had ever done before." Indeed he did, for Cook had his ears cut off.

This incident, following hard on the uncharacteristic violence on Moorea, raises questions. Was Cook becoming hard, losing his humaneness? Was his impatience a sign of extreme fatigue? Was it because he had been at sea almost continually for ten years and was now forty-nine years old? Historians have pondered these events, especially in the light of the circumstances surrounding Cook's death in Hawaii a year and a half later.

Omai was settled in his new house, and to help him and his two New Zealand companions, Cook gave him a horse and mare, a pregnant goat, an English boar and two sows, a full set of English armour, and a garden planted with a grapevine, a shaddock tree, pineapples and melons. Cook had now fulfilled his obligation to Omai as set forth in the Instructions from the Admiralty.

Only farewells remained. The description is Williamson's:

> Omai took his leave of us with a manly sorrow, until he came to Captn Cook, when wth all ye eloquence of sincerity he express'd his gratitude & burst into tears. The Captn who was extremely attentive to Omai y^e whole time of his being on board, and y^e pains he had take to settle him to his satisfaction in his native country, was much affected at this parting.

PLATE 21

A Young Woman of Otaheite [Tahiti], Dancing.

The Tahitian holiday was almost over, but there were still friends to be visited before Cook headed for the chilly northern seas. Some of his sailors became convinced that their destiny lay in sunny Tahiti and not in the frozen north. In Raiatea, where Cook stopped to visit his old friend Chief Oreo, and Captain Clerke delighted in once again meeting the chief's son, son-in-law and beautiful daughter, Poetua, the desertions began.

The first deserter, a marine named John Harrison, Cook personally tracked down. As Samwell tells it, the hapless youth was found "lying down between two Women with his Hair stuck full of Flowers & his Dress the same as that of the Indians." His punishment was two dozen lashes. A few days later two others made their escape and headed for Borabora in a stolen canoe. One of the men was the son of a prominent naval officer, and Cook felt some pressure to retrieve him, though this promised to be a difficult task. His plan was to take Poetua, her husband and her brother hostage and thereby convince Poetua's father, Oreo, that to gain his children's release he would have to put pressure on his own people to turn in the deserters. The abduction worked, though for several hours the *Discovery* was surrounded by "a most numerous Congregation of Women...cutting their Heads with Sharks Teeth and lamenting the Fate of the Prisners, in so melancholy a howl, as rendered the Ship...a most wretched Habitation." Clerke treated his captive friends with the utmost courtesy, and Poetua consented to Webber's request to let him paint her portrait.

Meanwhile, Oreo was furious and decided to retaliate by kidnapping Cook while he was taking his daily bath on shore. Receiving word of this plot from a Huahinean woman travelling on board the *Resolution,* Cook avoided capture, whereupon a foiled Oreo himself went to bring back the two deserters. The end of this affair "gave me more trouble and vexation than the Men were worth," Cook wrote, "which I would not have taken but...to save the Son of a brother officer from being lost to the World." Was Captain Cook softening a little with the attractions of these romantic islands? Would he really have permitted his sailors to desert under different circumstances?

The rest of the crew might well have speculated too, but Cook's capacity to bend the chief to his will must have dissuaded them from further attempts at jumping ship. On 9 December 1777, after a brief stop at Borabora, the Society Islands were left behind. Gilbert expressed the prevailing mood:

> We left these Islands with the greatest regret, immaginable; as supposing all the pleasures of the vayage to be now at an end: Having nothing to expect in future but excess of cold, Hunger, and every kind of hardship, and distress, attending a Sea life in general. and these voyages in particular, the Idea of which render'd us quite dejected.

PLATE 22

A View of Huaheine [Huahine, Omai's home].

Chapter Eight: Hawaii 1778

The Polynesian holiday was over: the real task of the voyage was to begin, and Cook had left himself plenty of time to make the passage to the coast of North America and then to reach 60° north latitude by June. Although he was ten months behind his original schedule, he was two months in advance of his revised one. Cook had never sailed the North Pacific before, but he knew that the wind system was similar to that of the South Pacific. He steered a course directly north with the southeast trade winds on his starboard beam, intending to work through the doldrums as best he could, pick up the northeast trades and sail north out of them. He was pioneering a route which later became widely used.

The two little ships kept close together on their passage through these unknown waters. Within two weeks they began to see birds, a sure sign that land was near. The Tahitians had given Cook to believe that no islands lay to the north of them, so it was with some interest that Cook approached the land sighted at dawn on Christmas Eve 1777. However, what they had discovered was only one of the many low-lying atolls that dot the Pacific Ocean. Cook named it Christmas Island and anchored there for a week to allow observation of a solar eclipse. Shore parties added some three hundred large turtles to the ships' menu, but there was no other food to be found, not even water, and the ships' provisions had to be used for the first time in the eight months since they had first anchored in the Friendly Islands.

For two seamen on the *Discovery*, this Christmas visit was a nightmare. Bartholomew Lowman and Thomas Tretcher nearly died after being lost for three days in the intense heat with nothing to drink except warm turtle blood. Cook commented wryly on their inability to find their way from one side of the tiny island to another: "A strange set of beings, the generality of seamen when on shore."

As they resumed the journey north, Cook issued heavy wool jackets and trousers to the crews as protection against deteriorating weather. He obviously did not expect to come across another tropical island, but on 18 January a sharp-eyed midshipman on the *Resolution*, James Ward, sighted land once more. The Hawaiian Islands had been reached for the first time by European vessels, and the discovery was to be of great strategic and commercial importance in the empty waters of the East Pacific. Cook named them the Sandwich Islands in honour of the Earl of Sandwich. For the ships' crews, who had not forgotten the pleasures of Tahiti, the islands held a serendipitous promise. As if to celebrate the discovery, grog was served for the first time since, on arrival in Tahiti some six months before, the crew had voluntarily agreed to forsake liquor and save it for the forthcoming cold journey to the north.

From a distance they could not tell whether the three islands that grew on the horizon were inhabited, but as they drew near and their high-flying sails were sighted by the natives, some canoes came off from the island of Kauai. Full of anticipation as to what new race of people they were now to encounter, the

English were amazed to hear these people speak what King described as a Tahitian "dialect." Cook and Clerke, two veteran sailors, began to compute in miles and degrees of longitude the staggering distances of wide open sea that separated these islanders from their kin in New Zealand, remote Easter Island, and all the archipelagoes that lay scattered within this triangle like so many random splashes on an immense oceanic canvas. Their astonishment increased as the ships hauled in close to the island and they could see people running abreast of them along the shore. More and more canoes came off bringing "roasting pigs and some very fine Potatoes, which they exchanged... for whatever was offered them."

The islanders' curiosity quickly overcame their fear of boarding the ships, and the first venturesome few were soon followed by great numbers. The English must have have been thoroughly delighted to witness their reactions. "I never saw Indians so much astonished at the entering a ship before," wrote Cook; "their eyes were continually flying from object to object, the wildness of their looks and actions fully express'd their surprise and astonishment at the several new objects before them and evinced that they never had been on board of a ship before." King gave them a tour of the *Resolution* and discovered that iron was familiar to them, though they did not know where it came from except to point to driftwood, indicating probably that they had pulled iron nails from floating timbers. "In their behaviour they were very fearful of giving offence," wrote King, "asking if they should sit down, or spit on the decks &c, and in all their conduct seemed to regard us as superior beings." If King were still smarting from the memory of smilingly superior Tongan dancers and Spanish-venerating Tahitians, this respectful treatment must have given him a measure of satisfaction.

Cook was immediately aware of the important practical implications of this first contact with an isolated island civilization. In fact, there was a serious purpose behind the enjoyable task of closely scrutinizing the Hawaiians and gauging their reactions to canvas sails, porcelain cups, iron tools, guns etc. For although the kudos that would be gained from their discovery of a new island civilization of such strategic commercial possibilities would be gratifying, Cook was also aware of the enormous risk of his sailors infecting the Hawaiians with venereal diseases. Therefore it was important to determine if the Spanish had happened on these islands on their Acapulco-Ladrona Islands trade route because, if they had, they would probably have introduced the diseases. Prevention was well nigh impossible, Cook wrote, since "where it is necessary to have a number of people on shore, the oppertunities and inducements to an intercourse between the sex, are too many to be guarded against." It became apparent during those first few hours of contact that the Hawaiians were not contaminated "with that greatest plague that ever the human Race was afflicted with." Some of the sailors, however, were suffering from it and Cook issued stringent injunctions: "In order to prevent [the disease] being

PLATE 23

An Inland View; In Atooi [Kauai].

PLATE 24

A Morai, in Atooi [Kauai].

communicated to these people, I gave orders that no Women, on any account whatever were to be admitted on board the Ships, I also forbade all manner of connection with them, and ordered that none of them who had the venereal upon them should go out of the ships."

Having set these limitations on their relations with the Hawaiians, Cook despatched "Three armed boats under the command of Lieutenant Williamson to look for a landing place and fresh water." When attempting to beach the boats, Williamson and his crew became apprehensive about the intentions of the crowd that encircled their craft. Some seemed to be offering assistance, others climbed into the boat and "a tall handsome man about 40 Years of age [who] seem'd to be a Chief," laid hold of the boat hook. As the boats rocked and almost tipped over, the crew were all for firing their muskets into the crowd, but Williamson held them back and attempted to gain control. "I went forward to the man who had hold of the boat hook, offering a nail, but he refus'd it." Williamson continued:

> I then made a stroke at him wth a small rifle barrel'd Gun which I had in my hand...then ...one of the natives made a stroke at me, at same time ye bow man call'd out he must let go the boat hook, I again turned about, & with ye greatest reluctance shot the indian who was struguling for the boat hook; it was then very apparent the good effect of these orders...the rest of ye natives...immediately quitted ye boat & fled to ye shore....

Williamson knew that Cook would disapprove, not only of the action but also of Williamson's avowed policy of not following warnings with threatening shots fired into the air. His shot had had the deliberate intent to wound or kill, but he did not report the incident to his Captain until they had left the islands and consequently Cook was given no opportunity to make the conciliatory gestures towards the Hawaiian people which he would have normally offered.

When the *Resolution* and the *Discovery* were securely anchored off the village of Waimea, the place recommended by Williamson, Cook decided to go ashore "to look at the water and try the desposition of the inhabitants, several hundreds of whom were assembled on a sandy beach before the village." It was late afternoon, many hours after news of the killing would have spread around the island. Cook recorded his experience:

> The very instant I leaped ashore, they all fell flat on their faces, and remained in that humble posture till I made signs to them to rise. They then brought a great many small pigs and gave us without regarding whether they got anything in return or no indeed the most of them were presented to me with plantain trees, in a ceremonious way....

This gesture was one generally reserved for the highest ranking Hawaiian chief. It was a mark of total submission, though in this case it may have been a gesture of appeasement. Cook was familiar with tabu-respect throughout Polynesia, but he had never been himself the object of such deference. The dividing line between high chief and divinity in Hawaii, as in most of Polynesia, was a vague one. On Cook's return to these islands in 1779 there was sufficient evidence to suggest that he was regarded as a "god-king." Were the Hawaiians, at this early point, already treating him as a divinity, even if only as a precautionary measure? Cook himself drew no conclusions. Typically he recorded only the facts and neither imputed motives to the Hawaiians nor indicated his own feelings on being given this extraordinary treatment.

That first night in Waimea Harbour, Cook permitted no one to sleep on shore, hoping by this further restriction to prevent venereal contagion. The next morning, 21 January 1777, he went ashore for the second time and was once again shown the highest homage. He sent some of his men to draw water from the pond behind the village with the help of some natives; then he set up an informal market on the beach in front of the village of Waimea. Webber's "Inland View of Atooi" (Kauai, Plate 23) almost lyrically captures the sense of peaceful harmony of the people and their setting.

Cook took a walk inland and upon his return found that the trade was brisk, orderly and very successful. "At sun set I brought every body on board, having got . . . Nine tons of water, and by exchanges chiefly for nails and pieces of iron, about sixty or eighty Pigs, a few Fowls, a quantity of potatoes and a few plantains and Tara roots." No people could trade with more honesty, he observed, ". . . never once attempting to cheat us, either ashore or along side the ships."

Cook's walk into the valley beyond Waimea took him past several villages and among taro, plantain, sugar cane and mulberry tree plantations. Accompanied by Anderson, Webber and a number of the Waimeans, Cook was everywhere greeted with the same remarkable compliment of full prostration. They noted that each village passed had a tall obelisk-like structure, and eventually they asked their guides to take them closer to one of the objects. They were led to a walled enclosure which in Hawaii is called a *heiau luakini* but which Cook called a *marae* because of its similarity to a Tahitian religious precinct. Webber's drawing of "A Morai in Atooi" (Marae in Kauai, Plate 24) evokes a heavy mood appropriate to feeling the presence of something dark and mysterious.

While inspecting the *heiau*, Cook was given to understand that important chiefs had been interred beside the *lananu'u* or divination tower, their remains being marked by roughly carved upright boards tied with strips of tapa-cloth. They also indicated where human sacrifices had been buried, one for each chief. The *heiau* had obviously been recently used for ceremonial purposes; fresh grass, which seemed to be used as an offering, had been placed before two

PLATE 25

The Inside of the House, in the Morai, in Atooi [Kauai].

female images in the enclosure's main building. Webber's drawing of "The Inside of the House, in the Morai, in Atooi" (Plate 25) is rather poorly rendered, the two female "godesses" taking on a definite masculine shape. It was not until his return visit to Hawaii in 1779 that Cook had an opportunity to observe the rituals associated with the *heiau*.

Cook's stay on Kauai was unexpectedly cut short by unsettled weather. The *Resolution* lost its anchorage while trying to manoeuvre a more secure position and was forced out to sea. Cook could not return to the anchorage because of swelling surfs and shifting winds. Clerke eventually followed in the *Discovery* and on 29 January 1778 they managed to find a suitable anchorage off the neighbouring island of Niihau. However, this brief sojourn off Niihau was to thwart Cook's attempts to prevent his men from leaving the "fowl disease" in the islands.

Cook sent a supply party ashore, intending to follow them shortly and survey this new island. However, once again the surf rose dangerously and not only was Cook prevented from landing but also the shore party was stranded for two days. Surveillance had been difficult enough on Kauai, where Cook knew that his orders had already been disregarded. To enforce them on Niihau proved impossible. Roberts's Log records for 25 January 1778, "Will Bradyley, for disobeying orders, with 2 dozen [lashes] and having connections with women knowing himself to be injured, with the Venereal disorder."

Cook finally managed to land on Niihau on 1 February 1778. He brought with him "a Ram goat and two Ewes, a Boar and Sow pig of the English breed, the seeds of Millous, Pumpkins and onions." These he put into the care of a Niihauan who appeared "to have some command over the others." (Cook had failed in his search for a chief or king on Kauai, although after the *Resolution* lost its anchorage Clerke was visited by a man who was obviously of very high rank.) Cook then took a short walk along the beach where at one point he was stopped by some of the inhabitants. The man who had the appearance of being a chief "began to mutter something like a prayer," reported Cook, "and the two men with the pigs [which Cook had brought ashore] continued to walk around me all the time, not less than ten or a dozen times before the other had finished." This ceremony, so laconically recorded and without any personal comment, was Cook's last experience of the Hawaiian Islands that year. That night, the anchor dragged and the *Resolution* again lost her position. Already frustrated by not finding good harbours on these islands, Cook decided to resume the journey to North America. For the second time they bade farewell to the salubrious clime of the Polynesian islands and headed for the chilly Arctic Sea. This time there was to be no reprieve.

The journal accounts of Hawaii in 1778 were understandably brief. On only three of the fourteen days spent there was it possible to land, and Cook himself spent no more than ten hours on shore. It was generally agreed by Cook, King, Clerke and Samwell that the Hawaiians bore a striking resemblance to the Tahitians, and this initial impression led the English to liken things to "that of Otaheiti" rather than to observe things in their own terms. The surgeon Anderson's comments would no doubt have proven interesting, but his failing health caused him to put aside his journal upon leaving Tahiti, and the last part of it was subsequently lost. Although King's Hawaiian chronicle is also short, he blossoms into speculative prose in Nootka Sound, perhaps under the surgeon's tutelage.

Chapter Nine: Nootka Sound

The departure from Hawaii for the coast of New Albion marked the beginning of the search for the Northwest Passage, the primary purpose of Cook's Third Voyage. Cook's instructions directed: "upon your arrival on the Coast of New Albion you are to put into the first convenient port to recruit your Wood and Water and procure Refreshments." Samuel Hearne's overland journey in 1771 had disproved Juan de Fuca's supposed inland sea route at 47° to 48° north latitude, and so Cook's Instructions directed him to then "proceed Northward along the Coast as far as the Latitude of 65°... taking care not to lose any time in exploring Rivers or Inlets, or upon any other account, until you get into the beforementioned Latitude of 65°, where we could wish you to arrive in the month of June...." Cook's route had been established on the basis of a map published in 1774 showing recent Russian discoveries in the North Pacific, amongst them an entrance to a passage to the Arctic Sea at 65° north latitude.

On 7 March 1778 Cook sighted the Oregon coast at 44° 13' north, thus becoming the first Englishman since Drake to have viewed both coasts of America. Two hundred years before, Drake had named this coast "New Albion" and claimed it for Elizabeth I in defiance of the pretensions of the King of Spain. It was still known only to a few brave Spanish seamen and an occasional Russian fur trader. The journals of Cook's officers reflect their curiosity as they tried to glimpse the shoreline through the haze. But it was too wintry and stormy for the ships to come in close.

The Oregon and Washington coasts became notorious for hazardous conditions in the days of sail; Cape Flattery, named by Cook, was considered one of the four most vicious headlands in the world. Westerly winds now put Cook's ships in great danger of being blown on shore. There were fogs, gales and appalling squalls. Cook was forced south as far as Cape Blanco before the winds turned in his favour and he could sail north. Even so, he had to stay well out to sea, and thereby missed the mouth of the Columbia River.

Towards evening on 20 March 1778 Cook stood off Cape Flattery, which he described as "well covered with wood and had a very pleasant and fertile appearance." Nightfall and a gale caused him to miss entirely the Strait of Juan de Fuca and he commented disdainfully in his journal, "It is in the very latitude we were now in where geographers have placed the pretended strait of Juan de Fuca, but we saw nothing like it, nor is there the least probability that iver any such thing exhisted."

On 29 March 1778, two weeks after arriving at the coast, the two ships entered a large bay which Cook called Hope Bay. They passed through a narrow channel into what he at first called King George's Sound (Nootka Sound) and finally anchored in Ship Cove (Resolution Cove) on Bligh Island.

As the ships entered Nootka Sound, thirty or forty canoes, each carrying two to seven men, approached and surrounded them. King described the encounter:

The first men that came woud not approach the ship very near and seemed to eye us with Astonishment, till the second boat came that had two men in it; the figure and actions of one of these were truly frightful; he workd himself into the highest frenzy, uttering something between a howl & a song, holding a rattle in each hand, which at intervals he laid down, taking handfuls of red Ocre & birds feathers & strewing them in the Sea; this was followed by a Violent way of talking... at the same time pointing to the Shore, yet we did not attribute this incantation to threatening or any ill intentions towards us; on the contrary they seem'd quite pleas'd with us; in all the other boats, someone or other act'd nearly the same as the first man did.

Although they realized that this was to be a "welcoming" ceremony, the English saw it mainly as a grotesque and wild display.

When the *Resolution* and the *Discovery* had anchored, the canoes came alongside, though none of their occupants would come aboard. The Indians made it known that they wanted iron, and this was traded immediately for some masks, bark cloth and bear skins.

All of the officers commented uniformly upon the appearance of the inhabitants of Nootka Sound: they were "a set of the dirtiest beings ever beheld," while Cook, on this first meeting, noted only their manners: "They seemed to be a mild inoffensive people ...more desirous of iron than anything else... which was traded with the strictest honisty on boath sides."

After securing a good anchorage in Ship Cove, Cook concerned himself with readying the ships for the northward journey. The severe buffeting experienced by the ships along the coast necessitated a considerable overhaul. They had to be re-rigged, caulked and generally made trim. Because the foremast of the *Resolution* was found to be rotten and the mizzenmast had collapsed, both had to be replaced. A blacksmith's tent was set up on shore while two trees were felled, stripped and hauled to the beach. Expecting the work to occupy them for one or two weeks, Cook had the observatory set up on shore on an elevated rock close by the *Resolution*.

The repair work allowed little time for leisure. The anticipated stay of two weeks stretched into four when the spar chosen for the mizzenmast was found to be defective, necessitating a new beginning. Work was kept at a grueling pace in order not to lose the season for searching for the Northwest Passage. Cook was preoccupied, and his observations of the surrounding country and its people were cursory in comparison to his Polynesian accounts. Friendly relations were established, but time and circumstances—especially the language barrier—did not permit the cultivation of friendships. Consequently, no personalities emerge out of this stay: no King Paulaho or "King" Tu.

All of the accounts of the inhabitants of Nootka Sound and, for that matter, of peoples encountered farther north as well, differ from the Polynesian ones in a number of respects. Nootka Sound was a harsh environment lacking the aesthetics and sensualities of a South Sea clime. Between the sultry Polynesian islands and the blustery shores of the northwest coast of North America, the crews of the *Resolution* and the *Discovery* had experienced a fundamental shift in attitude. The stormy coast of New Albion, coupled with the appearance of the Indians of Nootka Sound, had given their senses—in Clerke's words—"a hearty rattling." Furthermore, the inhabitants of the New World were already thought to be "Depraved Savages" rather than "Noble Savages." This contrast must have been heightened as the halcyon Polynesian days dissolved in rain and sleet and the sailors were confronted with people who looked wild and dirty. The Englishman's fastidiousness, with its attendant contempt, came easily to the fore in the presence of people who did not perceive dirt in the same way he did. The officers emphasized the contrast by constant comparisons between the Polynesian Islanders and the Nootka Sound Indians.

But Cook and his officers were careful observers and saw, for example, a people with a shrewd sense of how to trade well. Their sense of private property came as a surprise: the English had to pay for wood, water and grass for the ships' goats. The ethnographic portrait which Cook, King, Samwell and Clerke present in their journals is the first such record of the people of Nootka Sound. It is a remarkable achievement in the fact of Cook's admission that "we had learnt little more of their language than to ask the

PLATE 26

A Man of Nootka Sound.

names of things and the two simple words yes and no."

Cook spent the first three weeks of their four-week stay at Nootka supervising the work on the ships' rigging. During this time his dealings with the native people were limited to trading either aboard ship or in the shoreside work area close by. In such transactions Cook was ever alert to who was benefitting most from the exchanges of goods. "Trinkets for gold" describes the terms of trade that generally subsisted between white men and the North American Indian until well into the nineteenth century, and the Nootka Indians were little different from other Indians in their eagerness to obtain the white man's goods, but they were sufficiently adept at bartering that Cook observed, "these people got a greater middly and variety of things from us than any other people we had visited."

Nor were Nootka people interested in beads, baubles and coloured feathers, as Cook noted:

> Nothing would go down with them but metal and brass was now become their favourate, So that before we left the place, hardly a bit of brass was left in the Ship, except what was in the necessary instruments. Whole Suits of cloaths were striped of every button, Bureaus & ca of their furniture and Copper kettles, Tin canesters, Candle sticks & ca all went to wreck.

That the Indians were skilled in trading was no wonder, for there was a great deal of trading among the different groups of the northwest coast. Whereas formalized exchanges characterized trade in Polynesian society, theirs was closer to open bargaining. Shrewdness was valued, and unlike the Polynesians they did not complicate trading with tabus or sanctions which could benefit one trader more than another or restrict the number of people with whom one could trade. Cook and his men were not treated any differently; when they wanted something, they were asked to pay for it. Two incidents illustrate this rule.

When John Webber went into the village of Yuquot to do some sketches, he first drew a general view of the houses ranging the shoreline (Plate 27) and then sought to draw an interior scene "which would furnish one with sufficient matter to convey a perfect idea of the mode these people live in." He soon found a suitable house and proceeded to draw it (Plate 28).

> While I was employ'd a man approach'd me with a large knife in one hand seemingly displeas'd when he observ'd I notic'd two representations of human figures which were plac'd at one end of the appartment carv'd on a plank, and of a Gigantic proportion: and painted after their custom. However I proceeded, & took as little notice of him as possible, which to prevent he soon provided himself with a Mat, and plac'd it in such a manner as to hinder my having any further a sight of them. Being certain of no future opportunity to finish my Drawing & the object too interesting for leaving unfinish'd, I consid-

PLATE 27

A View of the Habitations in Nootka Sound.

ered a little bribery might have some effect, and accordingly made an offer of a button from my coat, which when of metal they are much pleas'd with, this instantly producd the desir'd effect, for the mat was remov'd and I left at liberty to proceed as before. scarcely had I seated myself and made a beginning, but he return'd & renewd his former practice, till I had disposed of my buttons, after which time I found no opposition in my further employment.

Cook's experience was almost identical, though what he wanted was grass for the ships' goats.

I went ashore and sent some people to cut grass not thinking that the Natives could or would have the least objection, but it proved otherways for the Moment our people began to cut they stoped them and told them they must Makook for it, that is first buy it. As soon as I heard of this I went to the place and found about a dozen men who all laid claim to some part of the grass which I purchased of them and as I thought liberty to cut where ever I pleased, but here again I was misstaken, for the liberal manner I had paid the first pretended pr[o] prietors brought more upon me and there was not a blade of grass that had not a seperated owner, so that I very soon emptied my pockets with purchasing, and when they found I had nothing more to give they let us cut where ever we pleased.

These experiences led Cook to comment that nowhere had he met Indians "who had such high notions of everything The Country produced being their exclusive property...."

Cook also observed that the Indians of Nootka Sound were establishing exclusive rights to the trade market with the English. Other Indians who approached Nootka Sound in order to trade were forced to deal through Nootka middlemen or not at all. When the visitors were allies, the Nootkans permitted direct trading, or they boosted the price of the trade goods in order to benefit their friends.

It became clear to Cook that there were internal divisions and rivalries between the people of the Sound. Trade was monopolized by a few powerful groups, and "the Weakest were frequently obliged to give way to the Strong, and were sometimes plundered of everything they had, without attempting to make the least resistance." Here were the rudiments of Nootka class structure. Had he more time to observe these people, Cook may well have produced a document on Nootka Sound that was as comprehensive as José Mariano Moziño's *Noticias de Nutka* of 1792.

On 20 April, three weeks after the ships' arrival in Nootka, most of the heavy work on the rigging was completed. Cook now felt free to explore his surroundings more closely, and so, with two boats and a crew of his midshipmen, he set out on a day-long expedition. They crossed the Sound to its western point and stopped in the village of Yuquot. Cook seemed to be known there, for he was received "very curti-

PLATE 28

The Inside of a House in Nootka Sound

ously, everyone pressing me to go into his house, or rather apartment for several families live under the same roof, and there spread a mat for me to sit down upon and shewed me every other mark of civility."

Both in the village and the houses Cook had an opportunity to observe the women for the first time. Unlike the "South Sea Island Girls who in general are impudent and loud," according to Samwell, the Nootka women "were very modest and timid," making no independent attempts to solicit the attentions of the sailors. The women, as well as the men, were "commonly grave and silent... by no means a talkative people." With a few rare exceptions (Plate 29), none of the English had anything complimentary to say about the physical appearance of the Indians. Captain Cook, who was the least likely to make critical comments of this kind, declared that "hardly one, even of the younger sort, had the least pretentions to being call'd beauties," while Samwell, who often dwelled on such descriptions, commented:

The Women were dirtier & more loathsome & despicable looking Figures than the Men, they were never cloathed in Skins or adorned with ornaments of any kind. Their Hair full of filth hung over their Faces in clots, in short a woman could not present herself to us under a more disagreeable form than these did....

This first visit to Yuquot was brief. Cook also wanted to survey the Sound, which he did, discovering that their anchorage was on a large island rather than the "continent." (Cook never discovered that Nootka Sound itself lay off an island. Having missed the Strait of Juan de Fuca, he believed he was viewing the coast of the North American continent.) The thirty-mile expedition which the rowers found arduous but agreeable because "Capt. Cook... would sometimes relax from his almost constant severity of disposition, & condescend now and then, to converse familiarly with us...," took them to another village opposite the northeast side of Bligh Island. Cook wrote:

The inhabitants were not so polite as those of the other I had visited; but this seemed in a great measure if not wholy owing to one Surly chief, who would not let me enter their houses, following me where ever I went and several times made signs for me to be gone; the presents I made him did not induce him to alter his behavour. Some young women, more polite than their surly Lord, dress'd themselves in a hurry in their best cloaths, got together and sung us a song which was far from being harsh or disagreeable.

One thing the English did agree upon was the pleasing quality of Nootka songs. They were rhythmic and complex, very melodic, and frequently expressed a melancholy which the English found engaging. Canoeists often sang in unison, keeping time by beating their paddles against the sides of the boat.

Colourful rattles in the shape of birds and filled with pebbles were used as rhythm instruments (Plate 30).

There were songs for all occasions, known to all and sung by everyone. But there were also songs that were the personal property of individuals or families; they could only be sung by members of that family and they were passed down through the generations like material property. Songs were not purely for pleasure or for entertainment but went beyond aesthetics, being first and foremost power possessions with great social significance. A family song, for example, proclaimed rank and prestige, while a Shaman's song, accompanied by the Raven rattle, had the power to lure evil away. But the subtleties of Nootka songs, like the social nuances conveyed in Tongan dances, were not perceived by the English.

After the expedition of 20 April, Cook made only one more visit to an Indian village before leaving Nootka Sound. Accompanied by Clerke and Webber, he again crossed over to Yuquot; here Webber sketched the interior of a Nootka house and the view of the village which appear as Plate 42 and Plate 41 in the journal. This was the occasion of the grass-cutting incident described above, but Cook made no further observations about the day's events.

In fact, Cook's Nootka journal betrays an uncharacteristic lack of curiosity. It is Samwell and King and Clerke who flesh out the account of Nootka Sound, for they recorded their responses to people and events, whereas Cook confined himself to general observations, except when he wrote of their departure:

PLATE 29

A Woman of Nootka Sound.

[a Chief] who had some time before attached himself to me was one of the last who left us, before he went I made him up a small present and in return he presented me with a Beaver skin of greater value, this occasioned me to make some addition to my present, on which he gave me the Beaver skin Cloak he had on, that I knew he set a value upon. And as I was desirous he should be no sufferer by his friendship and generosity to me, I made him a present of a New Broad Sword with a brass hilt which made him happy as a prince.

This was Cook the gentleman, the formal ambassador of the English Crown and the familiar personality known from the Polynesian journal. The chief was presumed to have been Maquinna, a prominent figure in the subsequent history of Nootka Sound.

The *Resolution* and the *Discovery* sailed out of Nootka Sound on 26 April 1778, carrying fresh supplies of wood and water and the valuable trade goods acquired from the Nootkans. There had been the usual brisk traffic in "curios," and the expedition brought back to Europe the first examples of the material culture of British Columbia Indians such as masks, weapons and bark capes. However, the best acquisitions of the English had been animal pelts, which in China and Europe would fetch a handsome price. Special worth was attached to the luxurious pelt of the sea otter, *Enhydra lutris* (Plate 31), as illustrated by this story about Midshipman Trevenen:

When the ships were in Nootka Sound, Trevenen, accidentally holding the rim of a broken metal buckle in his hand, observed a native with looks of admiring envy gazing on the lovely trinket. It was immediately accommodated as a bracelet to his arm, and an offer made to treat for it. The Indian thought he had made an advantageous bargin, when a very fine sea otters skin was accepted as an equivalent. This skin on the return of the ships to China, sold . . . for three hundred dollars. . . . its value was such as to enable Trevenen to purchase not only what he wanted to supply his own necessities, but silk gowns, fans, tea and other articles, which he brought home as presents to his sisters and friends.

The price these pelts fetched in China so far exceeded anyone's expectations that some members of the crew wanted to return to Nootka for more sea-otter pelts before heading home.

The publication of Cook's Third Voyage journals spread the news about the easy accessibility of this untapped wealth. Commercial expeditions were quickly launched, the first fur-trading vessels arrived in Nootka Sound in 1785, and by 1825 the sea otter was practically extinct.

In the early trading days, Spain and England

PLATE 30

Various Articles, at Nootka Sound.

came to the brink of war over Nootka Sound, each one seeking the trade rights for the Nootka area. Spain was eventually forced to relinquish her claims, and in 1789 Capt. George Vancouver, who had been a midshipman on Cook's Third Voyage, formally took possession of the fur-trading posts of Nootka Sound in the name of the British Crown.

Cook's departure from Nootka was an incautious one. He had noted a storm brewing on the western horizon but was anxious to continue the journey. Within a few hours the storm broke violently and by the early morning of 27 April had reached hurricane force. The *Resolution* sprang an alarming leak, and though the ship's pumps prevented what could have been a disastrous flooding, the damaged stern proved troublesome for the rest of the voyage. Cook headed out to sea rather than risk the danger of being swept onto a lee shore. Land was not sighted again until 1 May, near Cape Edgecumbe.

Following his instructions not to lose any time in exploring the coast until he reached 65° north latitude, where it was thought that a passage through to Hudson's Bay or Baffin Bay would most likely be, Cook passed and named Cross Sound and Cape Fairweather. By 10 May the ships were off Mount St. Elias, which they had first sighted from 100 miles away. It had been discovered and named in 1741 by Vitus Bering, a Danish explorer in the service of Russia, who had also discovered the strait named after him.

Russian exploration of the coast that Cook was now approaching had produced the wildly inaccurate maps which Cook carried with him and which were to cause him great frustration and loss of time as he attempted to reconcile them with the actual coast. Stahlin's map of Russian discoveries showed Alaska as a large island separated from the North American mainland by a passage to the Arctic Sea 15° east of Bering Strait. Cook was now looking for this passage.

PLATE 31

A Sea Otter.

Chapter Ten: Prince William Sound and Cook Inlet

Tempestuous weather kept the *Resolution* and the *Discovery* well away from the rugged northern coast and its offshore islands, each with its helmet of trees, that guard the continent like a phalanx of barbaric warriors. As they approached 60° north latitude the coastline began to veer to the west. The weather cleared and Cook took the ships to within ten to fifteen miles of the mainland, progress now slowed by light winds. Aboard ship, curiosity was very much heightened by the unexpected direction the coastline was assuming. "We now found the Coast to trend very much to the west inclining hardly anything to the North," Cook noted on 9 May, with Mount St. Elias visible before them. Three days later he passed the southwestern tip of Kayak Island only to be confronted with a coastline extending nearly east to west. To the west the land dipped south; to the east could be seen an inlet. Cook chose the latter direction, for if Stahlin's map possessed any degree of accuracy, the "Island of Alashka," separated from Stachtan Nitada or the Great Continent of North America by a broad passage, should lie in this immediate vicinity.

On both ships there was an atmosphere of intense excitement. "We have Dr. Matys [Stahlin's] map of the N°ern Archipelago constantly in our hands," King wrote, "expecting every opening to the N°ward will afford us an opportunity to separate the Continent, and enable us to reach the 65° of Lat^{de}. . . . We are kept in a constant suspense, every new point of land rising to the S°ward damps our hopes till they are again reviv'd by some fresh openings to the N°ward." Their tension increased as it became apparent that the whole aspect of the coast before them was entirely contrary to "the Modern Charts, founded upon the late Russian discoveries."

On 12 May they sighted two points of land separated by a large inlet. Reflecting the anticipation aboard ship, Gore christened the westerly point "Cape Hold with Hope" while Cook, adhering to his predictable heraldic nomenclature, distinguished it only as "Cape Hinchinbrook." "This being the first opening we had seen in the Land since we left King George's Sound [Nootka Sound]" wrote Samwell, "Our Hopes of a Passage were somewhat revived, especially as the entrance here is wide and it had at first a very promising appearance." Their expectations were thwarted, however, by a thick fog which enveloped them just as they entered the opening. Unable to see even a mile before them, they sought a harbour within the Cape and prepared to wait for the fog to lift. Periodic clearings gave them glimpses of a wide expanse of water with no land visible in a northwesterly direction. But confirmation would have to wait for three days.

A deflated Samwell surveyed a "Country all round having a very desolate and dreary appearance being almost entirely covered with Snow." But if they were disappointed in this inhospitable looking land, their curiosity was aroused, for it was inhabited. Two large canoes filled with men came and circled the ships. Their form of greeting was reminiscent of the Nootka Sound welcoming ceremony, and they also would not

PLATE 32

A View of Snug Corner Cove, in Prince William's Sound.

board the ships. Clerke gave one man a glass bowl and in return he was presented with a beautiful garment made from bird skins. Clerke took a liking to these new acquaintances, finding them "fine jolly full fac'd Fellows, abounding to all appearances in good living and content: they have very cheerful countenances, and in their conversation with each other there appears a good deal of repartee and laugh."

The English had little opportunity to further their contacts with these people, for the next morning the weather was clearer and Cook decided to move the ships to a sheltered harbour where the *Resolution*'s leak could be repaired. He headed north, into the deceptively wide gulf that is Prince William Sound, and found a secure anchorage at the entrance to Port Fidalgo, in a place he named Snug Corner Bay (Plate 32). Though fog and squalls again impeded their vision, they were still hopeful that their course would bring them to Stahlin's Passage. However, the discovery would have to wait. They were to remain in Snug Corner Bay for three days while the *Resolution* was careened and her damage tended to.

Meanwhile, there was much to engage the crews' curiosity, for once again the ships were surrounded by canoes carrying the people of the area. Cook and King were particularly intrigued. Both were struck by the enormous differences between the Prince William Sound people and those of Nootka Sound. Cook's curiosity returned to him, and King was full of provocative questions. The Prince William Sound people appeared to belong to a different race, but King's intellectual curiosity, prefiguring the concerns of twentieth-century anthropologists, went deeper than such surface observations:

> Their Dispositions and tempers from what we have relatd will be found to differ widely from any set of people we have seen; and One can but wish that either our undertaking had permittd, or weather favour'd the seeing more closely the Coast between this and King Georges Sound, and to have had connexions with natives in diff^t parts, in order to have seen where this striking difference in their Colour, in their Canoes, implements, and their turn of mind commences.

Cook's interest took him in a different direction. He became keen to identify which race these people belonged to. He had a copy of *The History of Greenland*, an illustrated book containing "a Description of the Country and its Inhabitants" written by a Moravian missionary, David Crantz, and published in 1767. Book in hand, Cook compared the Greenland "Esquimauxs" with the inhabitants of Prince William Sound. In every respect—in appearance, dress and technology—the two peoples were convincingly comparable, but Cook, meticulous and thorough in deciding when there was sufficient evidence to prove a point, reserved judgement. "These people are not of the same Nation as those who Inhabit King Georges Sound [Nootka Sound]," he could confidently conclude because of his first-hand observation that "both their language and features are widely different. . . ." But, as to the problem of whether these people were Eskimos:

> These [people of Prince William Sound] are small of stature, but thick set good looking people and from Crantz discription of the Greenlander, seem to bear some affinity to them. But as I never saw either a Greenlander or an Esquemaux, who are said to be of the same nation, I cannot be a sufficient judge. . . .

It is this brand of scientific empiricism—the "seeing-is-believing" variety—which made Cook a master mariner. He was not, to use Vancouver's deprecating phrase, a "theoretical navigator": he went out and literally mapped the Pacific Ocean. This active, calculating frame of mind dictated his conclusions about people as well. Cook was not an armchair anthropologist.

Cook's educated guess was probably accurate. Modern anthropologists have referred to Prince William Sound as a "terra incognita" because today's inhabitants appear to be different from those of Cook's time. However, it is almost certain that Cook met with the Chugach Eskimos, a very small subarctic branch of the southern Eskimos. Only eight tribes are known to have resided in the area, the Tatitlek peoples probably being closest to Cook's anchorage in Snug Corner Bay. The uncertainty in this identification lies in the fact that Prince William Sound is situated in the centre of what was presumed to have been an active trade route between Tanaina Athapaskans in Cook Inlet, Eyak Indians in the southern part of the Sound, and Tlingit Indians even farther south.

The first face-to-face encounters with the Prince William Sound people sent a ripple of surprise through the ships. One of the seamen called out in astonishment that he had found a man with two mouths. Although the men had seen for themselves that antipodean people were not constructed backwards, as Europeans commonly thought, it was easy to believe such an anomaly in this remote new world. The explanation for the man with two mouths turned out to be quite simple. "Some . . . men and women," wrote Cook, "have the under lip slit quite through horizontally, and so large as to admit the tongue which I have seen them thrust through. . . ." Ornaments were affixed in this opening (Plates 33, 34), though some of Cook's crew decided its main function was to scare their enemies, since the tongue protruding from this hole gave the face a ghastly appearance. The double-mouth effect was convincing enough that, according to Trevenen, "had we seen these people only once or slightly, we might very well have gone away in this idea, equally probable with those of Men with 2 heads, 1 Eye & etc."

These people shared with their neighbours in Nootka the same "thievish disposition," though they possessed none of the guile of the Nootkans. Rather, they proved the most audacious plunderers Cook

PLATE 33

A Man of Prince William's Sound.

had yet met with, leading him to conclude that they knew nothing of the power of muskets. They attempted to take over the *Discovery* in broad daylight and, when they were confronted by a horde of sailors armed with cutlasses, they simply returned to their canoes and headed over to the *Resolution*. There they attempted to take by force one of her boats. "But the instant they saw us preparing to oppose them they let her go steped out of her into their canoes and made signs to us to lay down our arms, and not only seemed but were as perfictly unconcerned as if they had done nothing amiss."

Aside from these incidents, goods generally changed hands by the more orderly methods of trade. Sea-otter pelts were the main item of interest for the English, while the natives wanted either iron or sky-blue glass beads. The iron pieces had to be large enough to be used for spears—eight to ten inches long and three inches wide—or they refused to sell their furs. The colour of the beads was also specific, like that of the trade feathers—always red—in Tahiti. The Prince William Sound people used the beads to decorate their hats (Plate 33) and as baubles to suspend from their numerous facial ornaments. According to Ellis, "for five or six of [these beads] a beaver-skin [sea otter] might be purchased worth ninety or a hundred dollars." The natives already had beads, but not many, and those they had were probably obtained by trade from other peoples who in turn got them from Russian traders operating out of the Aleutian Islands.

On 16 May the fog finally lifted. For the first time Cook was able to see that they were in a sound. The following morning, Stahlin's map in hand, the search continued. They passed numerous inlets but in each one the direction of the flood tide proved there could not possibly be a passage. There was no choice but to abandon Prince William Sound and take the southwesterly route.

Cook did not dismiss Stahlin's map just yet. A westerly course might still lead him to the supposed northern archipelago and the passage that would save him the long journey through the Bering Strait.

Tacking away from Prince William Sound and past the Kenai Peninsula, Cook made an important identification. Using Bering's and Stahlin's maps he saw a point of land which he was sure was Cape St. Hermogenes. On Stahlin's map this small island was situated almost due south of the passage they sought. "Everything inspired us with hopes of finding here a passage Northward without being obliged to proceed any further to the South," wrote Cook, while Gore christened this seemingly pivotal point in their journey, "Hope's Return."

The expedition headed into Cook Inlet (Gore named it the Gulf of Good Hope), a broad, deep waterway that cut a convincing northward gash into the continent. It was four days before the truth became apparent, and place names that reflected the dashed hopes of the officers appear in the journals and charts like accusing judgements against the land itself. They pass "Doubtful Island," and Gore's "Gulf

PLATE 34

A Woman of Prince William's Sound.

of Good Hope" becomes "Seduction River," its head designated "Turnagain Arm."

Frustrated, Cook took the expedition out of the inlet, with the statement that much valuable time had been wasted with nothing to show but "a trifling point in Geography." To a master hydrographer like Cook, there could be few trifling points in geography. And as if to underline his annoyance with inept mapmakers and "theoretical navigators" he comments, "But if I had not examined this place it would have been concluded, nay asserted that it communicated with the Sea to the North, or with one of these bays [i.e. Baffin or Hudson's] to the East."

Cook's final observation on leaving the area was that a very lucrative fur trade could easily be established with the inhabitants of the entire region, though Great Britain's participation was doubtful if a northern passage were not discovered. In these predictions he was quite correct. In Cook Inlet they had encountered more people, either Eskimos or Athapaskans, who had also offered furs in trade. Although the area had been sighted by Bering in 1741, it was not until Cook told the Russians he met at Unalaska of the wealth of furs that could be procured there that Cook Inlet and Prince William Sound became foci of Russian interest. When English and Spanish ships put into the Sound between 1786 and 1788, the Russian strength was already established and, by the end of the century, Russian fur trade companies held a monopoly.

Chapter Eleven: Unalaska to the Arctic Sea

Stahlin's map indicated that the shorter route to the Arctic Sea was to be found in a longitude approximately where Prince William Sound and Cook Inlet were situated, but in a latitude at least five degrees farther north. Cook was by now skeptical of these maps, but since the expedition had not yet reached 65° north latitude, he reserved judgement and followed the southwesterly trend of the coastline in search of an opening leading north.

The month of June saw the ships traversing the southern aspect of the Alaska Peninsula and Aleutian Archipelago. Barren, rugged hills, occasionally "ending in pointed rocks steep cliffs and other romantic appearances," emerged and then disappeared in confounding fogs and gales, making surveying and charting difficult.

Just off the Schumagin Islands, three or four canoes sped from shore and approached the *Discovery*. Clerke reported that "An Indian in one of them made many signs, took off his cap and bowed after the manner of Europeans, which induced them to throw him a rope, to which he fastened a small thin wood case or box and then, after speaking something and making some more signs dropped astern and left...." The English were certainly surprised, much more by the "European bow" than by the note that was found in the box. However, this indecipherable script, presumed to be Russian, posed intriguing questions. Had they been mistaken for Russians? Was the note from some stranded sailors awaiting rescue? The latter was unlikely, Cook insisted, since no foreigner was among the messengers, and so he continued his course.

They were now coming to the Aleutian Archipelago which stretches proboscis-like from the North American continent some nine hundred miles into the Bering Sea. The inhabitants of these islands, with whom the English were to shortly become more familiar, were known as the Aleut. They constituted the most southerly branch of the Aleut Eskimos, sharing with their distant Chugach neighbours the sleek, swift, sealskin covered *kayak* (Plate 37) and the same fondness for facial ornaments and Russian blue beads (Plates 35, 36). Also, like the Chugach, they were dependent upon the sea for their food and lived in coastal villages.

On sighting the most easterly of the Aleutians, Cook did not at first realize that it was an island. He missed the Unimak Pass, which would have admitted the ships to the Bering Sea, but on 25 June 1778 he sighted and made for a large opening in the land. With their approach came darkness and an enveloping fog, yet they sailed on until the sound of surf breaking on a shore was heard off the larboard bow and Cook immediately heaved anchor. The following morning a ripple of horror spread through the ships. Blindly in the night they had sailed between two rocks separated by only a quarter of a mile, a passage that even on a clear day, Cook wrote, he would never have ventured. The two rocks now stood behind them, Scylla and Charybdis, and Cook, perceiving his good fortune, called the island before him "Providence Island."

PLATE 35

A Woman of Oonalashka [Unalaska].

Today it is called Unalaska Island, in a pass named Unalga. The swift tides through the Pass created hazardous waters for ships to sail, and it was two days before both ships were through and anchored temporarily in Samgoonoodha Harbour off Unalaska. Before them they could see an open expanse of northward-trending waters. At last, they thought, their perseverance was being rewarded.

All that was needed at Samgoonoodha was fresh water and a supply of wood, but thick fog delayed departure for three days. This wait was not entirely unwelcome, for the English were curious to learn more about the inhabitants of the area, who bore a close resemblance to those encountered at Prince William Sound though they understood none of the words which the English had picked up from the Chugach people. Then there were their European manners of bowing and doffing their caps. And, as if these gestures were not startling enough, the Unalaskans offered no furs in trade and were not the least addicted to thieving.

These were the manners and attitudes that the eighteenth-century mind associated with the "civilized nations," and Cook, knowing the Russians to be familiar with this region, began to suspect that the Russian presence in the area was a formidable one. Besides the courteous restraint and respect for property, there were other unmistakeable signs of this civilizing influence. These Aleutian Eskimos had little to offer in trade other than fishing implements, especially their finely made harpoons with carved bone heads, but "they beg'd very hard for Tobaccy & wou'd pull of their hats upon receiving a Quid..." observed one of the Mates on the *Resolution,* and, from Captain Clerke, "their extreme avidity for Tobacco & Snuff is very extraordinary, they seem to think them the first of all blessings; but they would not touch any kind of spirituous Liquors."

These observations, and especially the fact that the Aleut offered no sea-otter furs in trade, suggested that the Russians were well established among the Aleut. Having spent only a short period of time there though, Cook refused to leap to conclusions; after all, he had not so much as seen a Russian, although the Aleut delivered numerous indecipherable Russian notes to him. But King was quite opinionated. He suspected that it was Russian policy to introduce the addictive tobacco as a means of securing trade relationships, and to price tobacco higher than the furs so that the Indians were always in debt to them. The tobacco, he wrote, had become "a necessary part of their subsistence; & the effects were visible, for the Natives were not only without Skins to barter, but it appeard evident that they could not afford any for their own clothing, which was almost entirely made of the Skins of Birds...."

King was not entirely critical of Russian "colonial" policy: "The whole of their [Aleut] behaviour gave us a good Opinion of their Masters who seem to have thought it worth the trouble to Correct their passions, & make them better members of society...." This was a significant opinion that Cook and

PLATE 36

A Man of Oonalashka [Unalaska].

PLATE 37

Canoes of Oonalashka [Unalaska].

Clerke shared. The Aleut's adoption of civilized manners was thought to be a sign of a qualitative improvement in them as a "nation." Although King was cognizant of the price that was paid for Civilization, namely, "there were many appearances of these People being abridg'd in their freedom, & against whom force must have been us'd... [and] it often appeard that they were fearful of selling us things, because they had not permission so to do," nevertheless the benefits, King thought, outweighed the costs.

Consequently, the English came away from Samgoonoodha Harbour on this first visit with a good impression of the Aleut. It is necessary to stress the brevity of this encounter, for they met only a few individuals and, with the exception of Samwell, none of the other journalists seem to have closely examined the Aleut houses or towns. Their opinions underwent a significant revision after a more lengthy return visit in October.

Webber's three engravings made during the first few days of July 1778 refer to notable persons and observations recorded in the officers' journals. Plate 35 "A Woman of Oonalashka," illustrates the respect with which the English wrote about the Aleut. Samwell and Cook each wrote of a woman; predictably, Samwell met "a very beautiful & truly good natured young Woman" while Cook mentions a middle-aged woman whom he observed with her family. They were "a very good looking people and decently clean" he noted in his log, and this woman in particular, who appeared to be the mistress of the family, was "of such a mien as would at all times and in all places command respect"—a rare kind of comment for Cook to make. Webber's Plate 35 could be either of these two women, though her clear countenance suggests the younger woman.

On 2 July the fog lifted from Unalga Pass and Cook, to everyone's delight, set a northeasterly course into the Bering Sea. Within a few days they were deceived by a deeply indented bay which Cook named Bristol Bay. "There never was a Map so unlike what it ought to be," wrote an exasperated King. The appearance of the land did not help, for it seemed "upon the whole a damn'd unhappy part of the World, for the Country appears just as destitute as a Country can be, and the surrounding Seas are scarcely navigable for the numberless Shoals, rising in different parts of them."

As they neared Bering Strait, the *Resolution*'s surgeon, William Anderson, died from a tuberculous infection he had carried with him from England. "The Death of This Gentleman, is a most unfortunate Stroke to our Expedition alltogether," wrote Clerke. "His distinguished Abilities as a Surgeon, & unbounded humanity, render'd him a most respectable and much esteemed Member of our Society; and the loss of his superior Knowledge... will leave a Void in the Voyage much to be regretted."

There was little to relieve the peculiar combination of stress and monotony that characterized these days, except perhaps a steaming pot of soup made from freshly plucked wild parsley, lupin root and the spinach-like lungwort gathered during a brief stop on Sledge Island on 5 August.

On 10 August 1778 Cook was off Cape Prince of Wales at the narrowest point of Bering Strait. Strong winds swept the ships towards the Asian (Siberian) shore and forced them to drop anchor in St. Lawrence Bay. "As we were standing into this place," observed Cook, "we perceived on the North shore an Indian Village, and some people whom the sight of the Ships seemed to have thrown into some confution or fear, as we could see some runing inland with burdthens on their backs."

Cook's account of the few hours spent with the Chuckchi is suffused with a probing intensity: he had to discover precisely who these people were. For the fact of the matter was that Cook was not entirely certain where he was. Was he still among Americans on the "Island of Alashka" and was the land behind him to the east the continent of North America? Or was he in the land of the Chuckchi on the Siberian coast with the Bering Strait and Alaska behind him?

Sensing that these people were hostile and were probably mistaking them for Russians, Cook approached them alone, unarmed and proferring small gifts. Cook's method of allaying a people's fears had been perfected by his vast and sometimes painful experience. He succeeded in gaining their confidence on this occasion; four or five of the men, who had been holding weapons in readiness, laid them down and performed a song and a dance.

A kind of bartering commenced and, as it did, Cook carefully scrutinized the strangers' possessions (Plate 38). Using David Cranz's book combined with his own observations of the Americans and his copy of Muhler's *Voyages From Asia and America* (1761), which included Vitus Bering's description of the Chuckchi, Cook made detailed comparisons but was forced to suspend judgement:

This land we supposed to be a part of the island of Alaschka laid down in Mr. Staehlins Map before quoted though from the figure of the Coast, the situation of the opposite coast of America and the longitude it appeared rather more probable to be the Country of the Tchuktschians explored by Behring in 1728. But to have admitted this at first sight I must have concluded Mr. Staehlins Map and account to be either exceeding erronious even in latitude or else a mere fiction, a Sentance I had no right to pass upon it without farther proof.

His restraint was admirable, though when he did prove incontrovertably that "Mr. Staehlins Map" was indeed erroneous, Cook did not mince words in asserting Stahlin's incompetence as a map-maker.

Cook's brief few hours spent with the Chuckchi had another quite different and unexpected outcome which Cook himself did not live to appreciate. A year later when the expedition, under Captain Clerke's command, reached Kamchatka en route to the second attempt to search for a passage through the Arctic, King picked up an interesting story. The Chuckchi, acting on the assumption that Cook was a Russian emissary, changed their attitude towards the Russians and had opted to pay voluntary tribute to them. Prior to Cook's visit the Chuckchi had vigorously resisted Russian territorial incursions. Czarist Russia under an expansionist policy had been spreading outward to Siberia since the sixteenth century, and this movement had intensified during the period of Cook's visit in 1778. Knowing of the bloodshed that had attended the Russian conquest of the Aleuts and the Kamchadals and contrasting it to Cook's confident but humanistic approach to the Chuckchi, King commented: "We felt no small satisfaction in having, though accidently, shewn the Russians, in this instance, the only true way of collecting tribute and extending their dominions."

When a favourable wind drove them through Bering Strait into the Arctic Sea, the question of where they were suddenly diminished before the vision of a northeasterly trending coastline. "All our Sanguine hopes begin to revive," wrote King, "& we already begin to compute the distance of our Situation from known parts of Baffins Bay." But all their hopes and computations were in vain. At 70° north latitude they spied on the northern horizon the brightness of reflecting ice, commonly called the "blink." Despite the persistent gloaming of this northern latitude, the significance of the blink did not register. It was not until a few hours later when a solid twelve-foot high wall of ice loomed massive and utterly impenetrable before them that they realized their search had come to an end. And it was an error-dispelling finale as well: neither of Cook's ships was equipped to work in ice because it was believed then that the sea could not freeze.

The drama of the search quickly became anticlimactic as excitement turned to disillusionment, then to fear. To be absolutely certain that there were no openings in the ice field, Cook spent a day cruising along its edge. Not only was there no broad opening but also enormous ice floes that were expanding and constantly shifting threatened every moment to thrust the ships into shoaling waters. Thick fog intensified the danger. For several hours they had to rely for directions on the roaring and braying of walruses which huddled in herds along the edge of the ice (Plate 39).

Extracting the ships from this icy danger meant veering sharply south and west, and Cook was forced to give the signal to retreat. When the dangers of the rapidly expanding ice wall seemed a little behind them, boiled walrus meat appeared on the mess tables, proving at first a highly palatable dish, especially after the monotonous daily fare of salt meat that they had endured during the cold journey north. However, the oily raw walrus flesh quickly turned rancid and within a few days was being compared to train oil. Some men stood staunchly by walrus meat as "pleasant and good eating...infinitely more nutritive and salutary than any salt provision" while

PLATE 38

The Tschuktschi [Chuckchi], and their Habitations.

others found it "disgustful, too rank both in smell and tast to make use of except with a plenty of peper and salt and these two articles were very scarce." Cook liked the walrus meat, and noted in his journal, "There were few on board who did not prefer them to our salt Meat." Young James Trevenen disagreed: "Capt. Cook speaks entirely from his own taste, which was, surely, the coursest that ever mortal was endued with." In fact, Cook had to flog two men for refusing to eat the fresh meat. A few days later, the spruce beer that had lasted eighty-two days ran out. It was bad timing, for the weather was bitter cold and the men had to work with rigging so encrusted with ice that their hands could scarcely grasp the ropes.

Meanwhile, Cook studied his maps and the ice formation they had encountered, to determine the possibility of the ice being navigable if they were to arrive during the peak of the summer season. Of one thing he had become convinced: "...that there is always a remaining store [of ice, i.e. it never melted completely] none who has been upon the spot will deny and none but Closet studdying Philosiphers will dispute." Nevertheless, he decided to return early in the following summer and in the meantime would head south along the same route they had come and explore more thoroughly the American coast to make quite certain they had not missed another passage. This decision was greeted with a "general joy" by a crew whose initiation to the Arctic had filled them with dread of being enclosed amongst the ice.

One last flicker of hope that they had discovered Stahlin's passage occurred as they entered Norton Sound, a large bay situated at about 64° north latitude that had been bypassed on the northbound journey. A close inspection revealed it to be another disappointing cul-de-sac. Cook decided to stop in the Sound for a few days to collect wood and water and to brew a fresh batch of spruce beer. They anchored in Chachtoole Bay, where they were visited by the Unaligmiut Eskimos who inhabited the Sound.

The English were by now quite familiar with the Eskimos. Said Cook: "None of us could perceive any difference either in the shape or features of these people and those we had met with on every other part of the coast King Georges Sound [Nootka Sound] excepted. Their cloathing which were mostly of Dear skins were made after the same fashion and they observed the custom of boring and fixing Ornaments to the under lip."

King, who was pursuing an independent line of speculation in his journal, had met a man and woman with their small child (Plate 40). The husband was blind and injured and, besides the whimpering child the woman carried on her back, there was a badly deformed young boy. The woman pressed King into performing what must have been a healing ceremony on her husband's eyes. King quite naturally appears to have been upset by this experience. He later wrote:

> The general appearance of these People seem to denote some Physical evil in their Origin, for their small size, dirty figure, sad countenance &

PLATE 39

Sea Horses [Walruses].

whining address, demands your commiseration, & shews that their situation is not so happy as that of their opposite neighbours, the Tchuskoi [Chuckchi]....

Here King was expressing the contemporary European image of the inhabitants of the New World, and his reflections were reminiscent of Comte de Buffon, who characterized the Americas as a harsh, cold, marshy, untamed world which stunted the growth of all species of life that were found there: a land where harmful insects bred in profusion but the human race degenerated into a small, sluggish, brutish species. King had first aired these thoughts when they met the Indians at Nootka Sound; a time when the journalists had this image of the "Ignoble Savage" juxtaposed with that of the "Noble Savage" of Polynesia, a lingering memory of Paradise Lost.

Upon leaving Norton Sound, Cook made a decision that delighted his crew:

Haveing now fully satisfied myself that Mr. Stahlin's Map must be erroneous and not mine it was high time to think of leaving these Northern parts, and to retire to some place to spend the Winter where I could procure refreshments for the people and a small supply of Provisions. ...No place was so conveniently within our reach...as Sandwich Islands [Hawaii], to these islands, therefore, I intended to proceed....

En route he stopped again at Samgoonoodha Harbour, where a deliciously seasoned salmon pie was delivered to the *Resolution,* which gift informed Cook that the Russians there were aware of the English presence and wished to make their acquaintance. A few days later, three Russian fur traders arrived from their small encampment in a harbour ten miles west of Samgoonoodha. King gave voice to a common sentiment:

The sight of these three Sailors rais'd a peculiar sensation in the breast of every one from the Captain downwards which was visible enough in our countenances, & our behaviour towards them. To see people in so strange a part of the World who had other ties than that of Common humanity to recommend them, was such a novelty, & pleasure, & gave such a turn to our Ideas & feelings as may be very easily imagined.

A few days later one Erasim Gregorioff Sin Ismayloff landed at the village where Cook, accompanied by Webber, was taking a walk. "He came in a Canoe carrying three people, attended by twenty or thirty other Canoes each conducted by one man," observed Cook. Ismayloff's servants set up camp for him and he invited Cook to his tent. "I felt no small Mortification," confessed Cook, "in not being able to converse with him any other way than by signs assisted by figures and other Characters which however was a very great help."

The contact with these Russians began to change the ideas and feelings of the English towards the Aleut who, from having been frequent visitors to the ships, now retired to the background. Cook, King and Clerke began to see the Aleut through Russian eyes, and as they observed and contemplated them at a distance, relations with this tamed, subject race became more formalized.

As might be anticipated, Samwell provided the exception. He was not interested in ice floes, maps, peculiar rock formations or people as familiar as Russians. His account of the three week stay at Samgoonoodha is rich with ethnographic details and characteristic racy anecdotes.

Samwell's nature was to experience things through participation, and he is at his descriptive best after he has been actively engaged with people. He had no queasiness about sampling what people ate, was not unduly repulsed by vermin or dirt, and had no compunction about exploring the sexual practices of the native people he met.

Aleut houses, he reported during the first visit, "are excessive nasty & stink most abominably of rotten Fish & other Filth which are suffer'd to lie about by the Inhabitants, they appear on the outside like a round Dunghill & the whole Town stinks worse than a Tanner's yard, occasioned by the Fish & whale blubber that is suffered to lie about in a state of Putrefaction" (Plates 42, 43). But Samwell's nose did not stop him and some of his mates from deciding to spend a night in one of these *barabras* or communal earth-houses, just "for the novelty of it." The women gave them some boiled shag (cormorant) for supper and, when the men arrived home from fishing, they offered the English their wives to enjoy, for Eskimo hospitality extends to all of a man's possessions—including his wife. After having given the men some tobacco, Samwell and company accepted their offer: "[and] so we pigged very lovingly together, the Husband lying close to his Wife & her Paramour & exercising such patience as would have done honour to a City Husband while we engrafted Antlers on his Head." Samwell should perhaps not be judged for his arrogance, since his intention is to be humorous, but it is difficult today not to condemn him for his often predatory attitude towards native women.

This particular episode fixed Samwell's attention on domestic details. He noted how they made fire using sulphur, how they tattooed with a needle rather than with striking instruments, and how they made baby cradles with skins stretched over a wooden frame shaped like a common coal scuttle.

Cook and King had mistakenly assumed from the Aleut's personal cleanliness and mannerly behaviour that cleanliness was a value that they would apply to every aspect of their lives, a wholly unwarranted deduction. Their discovery that this was not so not only lessened their previous high opinion of the Aleut but also made them less critical of the Russian methods of civilizing. In a comment that foreshadowed nineteenth-century colonial attitudes, King concluded his assessment:

PLATE 40

Inhabitants of Norton Sound, and their Habitations.

It is a subject perhaps as useless as it would be disagreeable, to draw a comparison between the [civilizing] Methods us'd... with nations in an unciviliz'd state, & to animadvert upon the effects of such treatment.... We are above instructing others in many things that seem to us of not sufficient importance to be worthy the trouble of teaching; & barbarous nations, being little more improv'd than Children, cannot comprehend higher matters. The truth is, we find that it answers our Selfish purposes better by falling in with their Customs & manners, than by attempting to make them confirm to ours.

The meeting with the Russian Ismayloff produced the final piece of evidence that persuaded Cook to totally discredit Stahlin's map. He now knew that he had stood on the Siberian coast and had met the Chuckchi; that there was no such "Island of Alashka"; that the only route to the Arctic Sea was via Bering Strait. His task for the coming summer was now cleared of the debris of inaccurate maps and the time-consuming surveying of unknown coasts. He would rest and refresh his crew and supplies in the Hawaiian Islands and, in the following spring, proceed directly to the Arctic Sea in the hopes of finding a navigable passage through the ice.

PLATE 41

Caps of the Natives of Oonalashka [Unalaska].

PLATE 42

Natives of Oonalashka [Unalaska], and their Habitations.

Chapter Twelve: Hawaii 1779

By the time the *Resolution* and the *Discovery* left Unalaska behind, the prospect of a winter in the relaxing warmth of the Hawaiian Islands was anticipated as a well-deserved holiday. It had been almost three years since they had left England, and still there was before them the task of a second sweep into the North Arctic. A winter layover in Hawaii would not only provide a welcome rest but also enable the ships to restock their supply of salt pork. With this latter thought in mind, Cook replenished his dwindling supply of trade iron at Unalaska by purchasing some from the Russians and converting it along with a damaged bower anchor into simple adzes. The previous year, three or four of these thirteen-by four-inch adzes had fetched a good-sized Hawaiian hog.

The journey south to the islands began with a fatality aboard the *Discovery* when, just out of Unalaska, a heavy storm broke. Gale-force winds rent and shredded sails and finally ripped the mast from its footing. The heavy timber collapsed, killing one man and severely injuring several others. When the fury abated, the battered ships resumed their journey, which was to take thirty tedious days. The crew members occupied themselves with mending the sails, repairing the small boats and making yarn.

On 26 November 1778, the island of Maui appeared on the horizon and grew steadily until the crowns of her high mountains could be seen vanishing into the clouds. As Cook hauled in, attempting to approach the island, he began to worry about finding a safe anchorage, for Maui's shores were rocky and steep and with a sea breaking against her in a thundering surf, she stood aloof like an island fortress. The ships' food supply was low but their water supply was ample, and seeing that some islanders were already coming alongside in canoes loaded with food for trade, Cook decided to ply along the coasts, standing offshore at night and encouraging a maritime trade during the day. Although such constant manoeuvring was taxing for his crew, Cook could be assured of a constant supply of fresh provisions and at the same time avoid making too great a demand on the limited resources of any one community. It was a prudent plan, considering that Cook anticipated remaining a few months among the islands, and it enabled him to look for a safe harbour in which to repair the ships. If Kauai, Niihau and now Maui were typical islands, then this search might well be a long one and require a good deal of patience and perseverance.

His plan was not received well by his crew. To ensure an adequate supply of food, Cook forbade all private trade for "curios" and placed control of all trading in the hands of his officers. He also ordered that no women be allowed on board, but after a few days this injunction was removed. Venereal disease was visible among the islanders, many of whom made it clear that they believed the affliction to have come from the English. But in spite of having access to women, the crew grew increasingly disgruntled. They voiced their complaints in a remonstrative letter to their captain, demanding more food and whiskey instead of sugar-cane beer. Cook was a little taken aback, but answered their demands with what he considered was a fair compromise. Since what they really wanted was to go ashore, the men remained dissatisfied, and took an obstinate stand of refusing to drink the distasteful decoction of sugar cane beer. Requests Cook would listen to, but insubordination he would not tolerate. Informing them that their actions were tantamount to mutiny, Cook refused to deal with them further.

Though it was certainly tedious and frustrating, the two months of cruising off the islands was not a lonely time. The islanders visited the ships in great numbers, boarding without hesitation, and always a few (inevitably some women) remained overnight, their canoes towed behind the ships. People on shore could often be seen waving white banners, a signal which was taken as a gesture of welcome and peace. Thievery was practically negligible, the majority of visitors proving to be exceptionally friendly, well-mannered and always careful not to give offence.

Nevertheless it was a "jaded & very heartily tir'd" company of Englishmen which finally, on 17 January 1779, brought the *Resolution* and the *Discovery* to anchor in the lovely Kealakekua Bay on the island of Hawaii. Their dampened spirits were revived by a welcome the likes of which they had never experienced. King estimated that they were surrounded by as many as fifteen hundred canoes. There were people standing on shore, people riding on surfboards, and several hundreds of women and young boys swimming around them "like shoals of fish." "I have no where in this Sea seen such a number of people assembled at one place," Cook wrote. King reckoned that there must have been ten thousand people present that day.

All decks were crammed with people who sang and danced and showed every sign of pleasure at their arrival. Their infectious joy spread to the crew even as they tried to control the crowds. First the *Discovery* and then the *Resolution* began to heel over dangerously with the weight of the crowds swarming the decks, but fortunately some proud young men dressed in rich feather cloaks and high-crested feather helmets (Plate 46) ordered the mob to jump overboard and averted a disaster. Two of these chiefs, Palea and Kanina, were ostentatiously authoritative, causing Cook and Clerke to single them out and to make their acquaintance. As it happened, they were kinsmen of the "King of Hawaii," a man called Kalani'opu'u who was presently in Maui. Palea and Kanina introduced Koa, an eminent elderly man who obviously "belonged to the the Church," as Cook put it. Koa, who subsequently assumed the title of "Brittanee," was the highest-ranking priest in the Kona district at that time. His superior, Kao, whom Lieutenant King designated the "Bishop," was attending Kalani'opu'u. Cook entertained them all for luncheon and later, at their request, he accompanied them ashore.

Until then, Cook's impression of the islanders was that they were exceptionally friendly. Their honesty and civility were apparent, and their respectfulness

PLATE 43

The Inside of a House, in Oonalashka [Unalaska].

PLATE 44

An Offering before Capt. Cook, in the Sandwich Islands [Hawaii].

sometimes bordered on awe. "It is very clear... from many... Circumstances, that they regard us as a set of beings infinitely their superiors," concluded King, but since King was susceptible to flattery, his first impressions cannot always be relied on. However, when Cook went ashore it appeared that, as in Kauai the previous year, something greater than "awe" and "respect" were being shown him. At a ceremony held near the *heiau* (Plate 24) the natives treated Cook in a way that, as King described it, approached adoration.

Captain Cook in fact became the central figure in two elaborate rituals performed by Koa and a "tall, grave young man named Kili'ikea," called the "Curate" by King (Plate 44). The captain, long distinguished by the islanders with the name of "Lono," had been presented before the carved images in the *heiau,* then led to the top of the oracle tower, where Koa and Keli'ikea performed a prophesying ceremony using the decomposed remains of a pig offering. First Koa picked up the "corrupt'd hog" and repeated a prayer over it. Then the two priests "kept repeating sentences in concert & alternately, & many times appeard to be interrogating; at last Koah let the hog fall, & he & the Captain descended." Cook was then asked to kiss the image of the god Ku, and with all of this Cook complied, or rather passively underwent, as King recorded it. No one was certain at first of the significance of these rites, only that it was highly respectful and seemed "to promise us every assistance they could afford." Unfortunately Cook left no record of his impressions or understanding of these ceremonies. His journal ended 18 January 1779, the day of arrival at Kealakekua Bay.

King is almost the sole source of information concerning the events leading up to their first departure on 4 February. He forged a respectful and seemingly intimate acquaintance with the priests whose houses stood near where the observatory tents had been erected. King lived on shore in the frequent company of the priests, while Cook remained on the *Resolution,* presumably attending to the chiefs who visited him daily. Cook's observations would have proven interesting beside King's because Hawaii, unlike Tahiti or Tongatapu, possessed a particularly powerful, high-ranking priesthood, something that King began to sense but not fully fathom. Nor did he, nor Captain Clerke for that matter, begin to suspect the possible implications of Cook's being designated "Lono" by the priests.

The ritual context and procedure that surrounded the offerings made to "Lono" (Plate 44) should surely have been evocative enough to prompt King and Clerke to look for the significance of the seemingly honorific title of Lono. King recorded one ceremony quite fully:

Kirikeeah [Keli'ikea]... with his face towards the Capt[n]... kept repeating in a very quick tone some speeches or prayers, to which the rest responded, his part became shorter and shorter, till at last he repeat'd only two or three words at

PLATE 45

Tereoboo [Kalani'opu'u], King of Owyhee [Hawaii], bringing Presents to Capt. Cook.

a time & was answerd by the Croud repeating the Word Erono [Ye, Lono]... which I suppose lastd near a Quarter of an hour....

Perhaps because they were accustomed to the lavish hospitality of "kings" like Finau, Paulaho and Tu, Clerke and King could not be expected to find King Kalani'opu'u's gifts to Captain Cook at all unusual. Kalani'opu'u had been much spoken of and the English waited with anticipation for the great king to appear. He arrived at Kealakekua Bay on 25 January 1779. The following day the entire harbour was declared *kapu* (tabu) so that the three great canoes carrying Kalani'opu'u and his attendants would have exclusive use of the sea (Plate 45). The king and his closest companions, dressed in their magnificent feather caps and capes, were regally ensconced in the largest double canoe. In the second sat the high priest, Kao, in charge of the three large feather-gods. The third canoe carried hogs, vegetables and fruit.

Lieutenant King turned out the marine guard to receive them on shore. Kalani'opu'u was a tall, thin old man who, in spite of bearing all the marks of a great *kava* drinker, carried himself with a dignified air. He greeted Captain Cook on the observatory marquee "& threw in a graceful manner over the Capt[ns] Shoulders the Cloak he himself wore, & put a feathered Cap upon his head, & a very handsome fly flap in his hand: besides which he laid down at the Captains feet 5 or 6 Cloaks more, all very beautiful, & to them of the greatest Value."

The feather cloaks were the pride and glory of Hawaiian high chiefs, who alone were permitted to wear them. They were made of hundreds of thousands of brightly-coloured feathers taken from rare birds and attached to a pliable net backing. These cloaks were symbols of divinity (Plate 46). A Hawaiian king or paramount chief (a more accurate designation) was linked genealogically to the gods: a human being, he yet participated in divinity. And thus it was that Cook was acclaimed "Lono" not only by impressive priestly rites but also by the power and might of an undisputed head of state.

It is difficult to know why Kalani'opu'u and his priests deferred so much to Cook. Kalani'opu'u was in 1779 seeking authority over the neighbouring island of Maui, but there seems little likelihood that he sought the aid of the English, for unlike Tu, Kalani'opu'u could not have been acquainted with the power of English weaponry.

On the other hand, there is some convincing evidence that the Hawaiians did link Cook with the god Lono. A Hawaiian tradition has come down from the eighteenth century that the islanders perceived the remarkable similarity between the sails on Cook's ships and the central symbol of Lono: the Lono-banner, which was a long piece of white tapa-cloth affixed to a T-shaped frame. Every year during the Makahiki season (October to January) which was sacred to the agricultural god Lono, the Hawaiians would construct a Lono-banner and carry it in procession at harvest time around the island, collecting offerings of goods and valuables for Lono. This tribute constituted a tax towards the support of the paramount chief, Lono's earthly representative. Coincidentally, Cook's two visits to Hawaii in January 1778 and November to February 1779 fell within the Makahiki season. Both times his white banner-like sails could be seen parading along the coasts, and as the islanders went out to greet him they were given precious metal tools in return for hogs, fruit and vegetables.

If Cook were linked with Lono in some fundamental way, this association must have eroded over time or with changing circumstances. The two and a half weeks spent at Kealakekua Bay constituted the most peaceful days of the entire expedition. Whatever apprehensions the English might have had in consequence of some thefts when they first arrived were soon dispelled. King voiced no reservations about how trustworthy these people were, saying that "What praise soever we may bestow on our Otaheiti friends & still more on those at the friendly Islands, we must nevertheless own, that we durst never trust them with such entire confidence as we have done these people." King often fondly remembered the boisterous, back-slapping Tongans and wished that the Hawaiians were not so cringing in showing their respect. On the other hand in friendly Tonga it was unsafe to walk alone in the countryside for fear of losing all you possessed to "lawless rascals," recalled Clerke: for example, Cook's servant, William Collett, had on Elysian Eua been set upon and robbed of everything except

PLATE 46

A Man of the Sandwich [Hawaiian] Islands, with his Helmet.

his shoes. By contrast, in Hawaii the gentlemen were entirely freed of such fears and ceased carrying their weapons.

What happened to later transform this profusion of hospitality into unfriendliness it is difficult to say, unless the cause lay in that very excess. The ships departed on 4 February 1779 carrying a present from Kalani'opu'u the value of which "exceeded any thing given at either the Friendly or Society Islands." Clerke voiced the sentiments of everyone when he declared that the Hawaiians' "extraordinary attention and friendship to us during our first visit was I believe without all precedent... here the face of insult was never seen; cordial attachment seemed to dictate every act throughout all ranks of people...."

They left this hospitable land with a view to seeing some of the other islands in the group prior to steering for Kamchatka and the Arctic Sea, but by this point in the voyage, trouble with the ships' rigging was almost a predictable occurrence. The *Resolution* was crippled when its mast was sprung during a storm just three days out of Kealakekua Bay. Cook's choice was not an easy one: either look for another harbour among these treacherous, fortressed islands, or return to the safe anchorage at Kealakekua where, Cook felt, they had drained the resources of the people. Vexed and somewhat uneasy about imposing upon the Kealakekuans once again, Cook nevertheless chose the latter plan, arriving back at the harbour on 10 February in time to celebrate the third anniversary of the beginning of the voyage.

A strange silence greeted them: the once-thronging Bay remained deserted. However, their fears were assuaged when they learned that Kalani'opu'u had just left and the harbour was kapu on that account. He returned the following day and visited the ships. Burney reports that the king "was very inquisitive, as were several of the Owhyhe Chiefs, to know the reason of our return, and appeared much dissatisfied at it." On the morning of the 13th the mast was taken ashore, the observatory tents were erected in their old place in the *heiau,* and a man was sent to fill the water casks at the well. This last activity produced some unexpected trouble. The man hired some of the islanders to assist him, but a chief prevented the people from helping. King sent a marine, whereupon the crowd collecting around the well armed themselves with stones. No stones were thrown, though some insults were aimed. As a precaution, Cook ordered the guards to load their muskets with ball, the lethal ammunition, instead of shot.

Later that afternoon, the armourer's tongs were stolen from the *Discovery* and the young chief Palea, who was on board at the time, took to his canoe in pursuit of the thief, along with some of the *Discovery*'s men in their pinnace. They retrieved the tongs but the thief got away. This annoyed Edgar and Vancouver, who seized the nearest canoe, which happened to be Palea's, intending to ransom it in return for having the thief delivered to their hands. Palea took offence and a fight broke out in which each side bashed the other with oars and paddles, neither Edgar nor Vancouver being armed. People came to assist Palea and the Englishmen were forced to swim for safety amidst a shower of stones.

This unfortunate brawl put everyone on edge. At dawn the following morning, 14 February 1779, Captain Clerke discovered that his large cutter had been cut from her moorings and was nowhere to be found. This was a serious theft, and on being informed of it, Captain Cook became furious. His actions to recover the craft were to cost him his life.

Accompanied by two officers and nine marines he travelled across the Bay at seven o'clock in the morning to take old King Kalani'opu'u hostage. He despatched armed boats to the entrance points of the Bay to ensure that no one escaped with the boat. The king was still asleep. When the lieutenant of the marines roused him and informed him that Cook was awaiting his presence to invite him aboard the *Resolution,* the elderly man, though still not fully awake, immediately agreed to come. As he met Cook on the beach, a crowd began to collect and the king's wife hurried to her husband, imploring him not to go because something seemed amiss.

Kalani'opu'u hesitated. An enormous crowd had accumulated out of nowhere and a great restlessness surged through them. Perhaps there was something ominous about the early hour, the presence of armed marines, the pathetic uncertainty of the bibulous old king and the warnings of the "queen" that made the people arm themselves with stones.

Just then a canoe sped over to the beach and a

PLATE 47

A Man of the Sandwich [Hawaiian] Islands, Dancing.

PLATE 48

A View of Karakakooa [Kealakekua], in Owyhee [Hawaii].

man called out that a favourite chief had been killed by Cook's men. Cook hesitated, uncertain of what was being said. A man lunged threateningly at him. Cook aimed at him and fired his musket, which was only filled with small shot. This act infuriated and emboldened the crowd, who worried for the life of their king. The crowd attacked, and in the fray Cook and four of his men were killed.

The news of Cook's death, so sudden and without warning, stunned the crew and they reacted with panic, fearing a general attack. The shore party and the half-completed mast were hastily brought on board amidst a confusion of stone-throwing by the Hawaiians and cannon shots from the *Discovery*. When all was gathered aboard, the two ships sat like tiny island fortresses in a large, empty harbour.

Within hours, shock at Cook's death was replaced by a rage for revenge, Clerke wanting to take "a stout party on shore, make what distruction among them I could, then burn the Town, Canoes &c." But he quickly realized that this would result in killing on both sides "and as the loss of a very few Men would now be most severely felt by us I thought it would be improper and probably injurious to the expedition to risk farther loss of the People." He gave orders that the ships were to be prepared at all times to launch an attack should the natives "not conduct themselves with some degree of propriety," and gave top priority to the repair work which forced them to remain in this hostile place.

On the evening of the 15th, Clerke sent Lieutenent King to within talking distance but still offshore to ask his friends the priests for Cook's body, which had not been recovered during the fray. Koa informed him that the body had been "cut up and carried away, but that he would try to get it for them." The following evening a portion of Cook's thigh was delivered to Clerke, who responded that the "sight [was] so horribly shocking, distraction was in every mind, and revenge the result of all."

The next afternoon, the 17th of February, a man appeared in a canoe wearing Captain Cook's hat. He flung stones at the ship while people on shore shouted and laughed. The insults being unbearable, Clerke fired some four-pounders into the crowd; tempers on board ship smouldered. A watering party sent on shore were harassed with stones, and the crew turned on their assailants with pent-up fury, shooting them, burning down houses, and, in a fit of barbaric vengeance, cutting off the heads of their dead victims and impaling them on stakes near the *heiau*.

With this violence the immense emotions began to abate. The islanders came to the beach carrying white flags and gifts of food. They had lost twenty-five of their own people, including the young chief Kanina, but seemed intent on mending the "miserable Breach" that was between them. The market was resumed, the chiefs once again came on board without their weapons, and Kalani'opu'u sent them a present of hogs. The last horripilating incident occurred when the bones of Captain Cook were finally returned six days after his death. On 22 February "at 5 [p.m.]

PLATE 49

A Canoe of the Sandwich [Hawaiian] Islands, the Rowers Masked.

both Ships hoisted Ensign's & Pendants half Staff up & Crossd over Yards, at ¾ past the Resolution toll'd her bell & fir'd 10 four pounders half Minute Gun's & committed the bones of Capt. Cook to the Deep, at 6 squar'd our Yards."

Captain Clerke took over the command of the *Resolution* and appointed John Gore captain of the *Discovery*.

There was no reason to remain any longer on Hawaii. The defective rigging had been repaired, and even though Clerke began to feel that once again he "might put some degree of confidence in [the islanders] with great safety," he still had to make a stop at Kauai to get water before heading for Kamchatka.

This he did, leaving Kealakekua on 23 February 1779 and on 1 March arriving at Kauai, or Hog Island, as Clerke called it on account of the extraordinary plumpness of the pigs to be found there. News of the events on Hawaii had preceded them and some initial scuffles resulted in the death of a Kauaiian, but for the most part the nine days spent there were peaceful. The politics of the island were significantly different from the previous year. A woman named Kamakaheli had wrested control of Kauai from Kaneoneo, who had been in power in 1778. Clerke attended to "Her Majesty" and enjoyed learning of her manipulations and exploits in the course of winning power.

In spite of the wintry prospect which lay ahead of the expedition, Clerke had no problems with desertion for, as he himself observed, "the idea of turning Indian which was once so prevalent among them as to give us a great deal of trouble is now quite subsided, and you could not inflict a greater punishment upon those who were the warmest advocates for this curious innovation in life than to oblige them to take that step which, 16 or 18 Months ago seemed to be the ultimate wish in their Hearts." Even the blandishments of Samwell's Hawaiian "Venus's (Plate 50) were insufficient to overcome the suspicion and horror in which they now held the islanders because of their captain's death. Although they knew that the Hawaiians had treated Cook's body according to the sacred rites reserved for high chiefs, no Englishman with a Christian upbringing could be expected to respond to the ceremonial mutilation of the human body with anything but revulsion. To remain amongst the islanders would have required submitting to the authority of the chiefs, which was unthinkable to most. The English had become strangers in paradise and so were glad to leave Hawaii.

The journalists completed their accounts of the Hawaiian Islands dispassionately with few self-accusations concerning their own gullibility or even their about-face reassessments of the Hawaiian character. Cook's death seemed to defy explanation or accusation.

John Webber's contribution to the Hawaiian chronicle gives no hint of a personal response to the tragedy; in fact, his drawings possess a lyrical, even fanciful quality. In "A Man of the Sandwich Islands,

Dancing" (Plate 47), for example, there is a truly elegant distortion of the human body, its magnificent proportions being the more heightened by a diminished background. "A Woman of the Sandwich Islands" (Plate 50) was definitely a figment of the engraver's imagination, for Webber's original sketch shows a rather coarse-looking young woman. "A Man of the Sandwich Islands, in a Mask" and "A Canoe of the Sandwich Islands, the Rowers Masked" (Plates 51, 49) are particularly enigmatic. Neither King nor Samwell, in their thorough accounts of island manufactures and articles of dress, mentions this gourd-like "helmet" adorned with foliage and tapa-cloth tassels, nor do they describe a scene in which the masked rowers appear. Considering the care with which the illustrations for the published journal of the Third Voyage were chosen, and the fact that Webber's were the first drawings made of these islands, it is puzzling to discover that four of the nine Hawaiian engravings stand alone without references to their subjects in the text.

PLATE 50

A Young Woman of the Sandwich [Hawaiian] Islands.

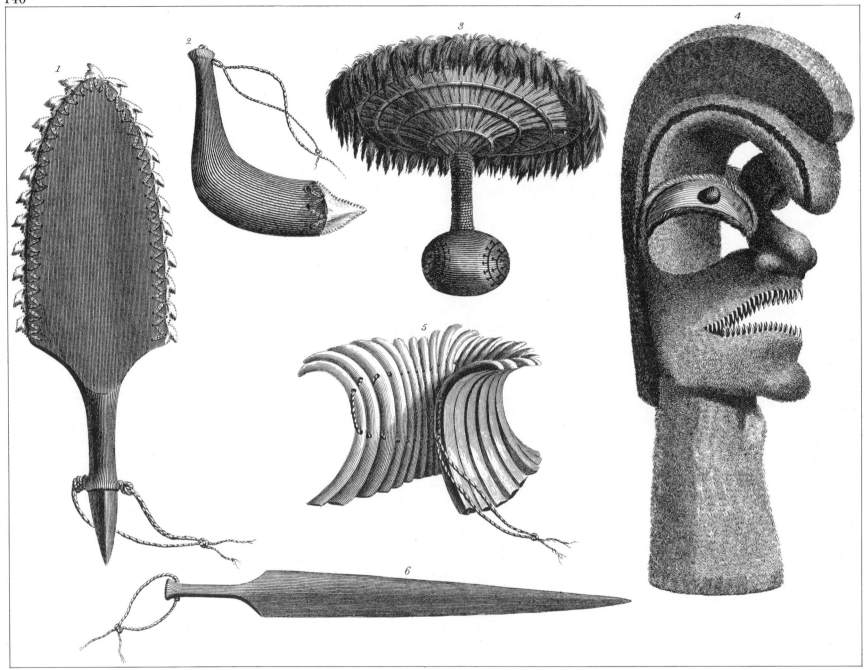

PLATE 51

Various Articles, at the Sandwich [Hawaiian] Islands.

PLATE 52

A Man of the Sandwich [Hawaiian] Islands, in a Mask.

Chapter Thirteen: Kamchatka and the Second Arctic Sweep

Within a month of leaving the Hawaiian Islands Clerke's men were experiencing temperatures as much as 53 degrees F. below Kauai's. It was just below freezing: weather which, Clerke remarked, "as matters go in general can by no means be call'd cold Weather but the tumbling so hastily from ye territories of the Sun into those of Frost & Snow of which we have now abundance gives us confounded chilly sensations."

Though Clerke never mentions it, his health was rapidly deteriorating and he spent much time in his cabin. But he tended to the welfare of his men by issuing heavy wool jackets and trousers to the "People whose Galantry among the Isles has render'd as naked as when born." However, there was not enough of this gear, which had served them so well the previous season, to provide everyone adequately. The ships pitched and rolled violently in heavy seas and the extra ballast taken on in Hawaii in addition to the necessity of running under full sail against gale-force winds caused the ships to strain and leak uncomfortably.

Upon leaving the Hawaiian Islands, Clerke had set them on a new track for the passage to Kamchatka, one that put them "still in the way of discovery." But all they sighted through the drenching fogs in that latitude were boobies, noddies and man-o'-war birds. As they moved steadily north, driving sleet turned to blankets of snow which froze on the sails and rigging. The ice-encrusted shrouds thickened into enormous blocks, and the two ships sailed upon a blackening sea like Arctic ghosts. The doctor's list of patients increased daily. On the morning of 24 April 1779, land was sighted beyond a lifting fog, and "a more dreary prospect I never yet came in the way of," wrote Clerke. "Every atom of Earth was cover'd with Snow except the sides of some Hills which...appear'd black Cliffs." They approached their destination two days later but the severe weather kept them outside Avatcha Bay, where the town of Petrapavlovsk was situated.

On that same day, 26 April, the *Resolution*'s faithful chronometer stopped. This was the precision timepiece that accompanied Cook on his Second Voyage and had performed with admirable accuracy ever since. Benjamin Lyon, an apprentice watchmaker on the *Resolution,* came to Captain Clerke's chilly cabin several times to clean and adjust the instrument, but it no longer worked.

The purpose of their visit to Petrapavlovsk was to obtain the beef and flour which the Russians in Unalaska had led them to believe would be readily available there. "Imagination could not paint a more dreary prospect," opined King as they entered the ice-bound harbour of Avatcha. "We saw a few log houses & Conic shaped huts upon Poles, but their miserable appearance & smallness in number would not allow us to take it for St Peter & St Pauls Village [Petrapavlovsk]." The British Admiralty had suggested in their Instructions to Captain Cook that he winter in this silent, solemn place, the very idea of which, wrote King, "made us Shudder again."

Having resigned himself to the fact that this was indeed Petrapavlovsk, Clerke sent King and Webber, who between them could communicate in French and German, to meet the Russian officers residing there. They were landed on the ice some distance from the village. The footing in the melting snow was treacherous, but they soon saw a dog sledge approaching them. "Whilst we were commending their ability in sending a Carriage to us, he turned short round & went great speed to the Ostrog [village]" (Plate 54). This reconnoitering occurred several times "attended with a few hearty damns from our party" and especially from a soaked and shivering Lieutenant King, who had fallen through the thin ice.

The Russian in charge of the village, Sergeant Surgutsky, received them civilly but with a restraint backed up by a military show of arms. Webber's German was once again useless, though King, with some hope, handed the Sergeant the letters of introduction obtained from Ismayloff at Unalaska the previous year. They tried to acquaint the Sergeant with their provisioning requirements, whereupon he indicated that he needed approval from the Governor of Kamchatka before he could sell them anything. The Governor, whose name was Major Behm, lived in Bolsheretsk, a small village close to the east coast of the Kamchatka peninsula. Sergeant Surgutsky sent Ismayloff's letters off to Major Behm while the English, confined to the ships because of the dangerous thawing ice, awaited the Governor's response.

Although Sergeant Surgutsky's hospitality on that first day extended to an excellent ten-course meal which included "fresh Beef Baked and Boiled. A Brace of Baked wild Turkeys. Fish cooked several ways and Pasty made of very good Flower," and all well lubricated with a bottle of brandy from the *Resolution,* King did not fail to notice that "two small field pieces were at the entrance to his [Surgutsky's] house, & point'd towards our boats, Shot, Powder & light'd matches all at hand."

It was several days later before the English began to suspect that some misapprehension lay behind these precautionary measures. Emissaries arriving from Major Behm displayed great surprise upon seeing the English vessels, and demanded that two English hostages be left in the village before they would board the ships. They delivered a "merely complimentary" letter from Major Behm "inviting Captn Clerke & his Officers to Bolchoireeka [Bolsheretsk]." One of them, a man named Port, spoke German and, while conversing with Webber, he indicated that Ismayloff's letters were to blame for their suspicious treatment by the Russians. As Ellis put it, Ismayloff's letter "had represented the ships only as packet-boats; that there were no officers on board either, and that he looked upon us in no better light than a set of sharpers, and that they would do well to be on their guard." With a view to correcting this misrepresentation and smoothing the way to obtaining the needed supplies. Clerke despatched King, Webber and Gore to Bolsheretsk to meet the Governor.

PLATE 53

Summer and Winter Habitations, in Kamtschatka [Kamchatka].

Port intimated to the English that their arrival had caused a flurry of fear in this Siberian outpost of Russian civilization. Since none of the villagers was familiar with European languages, they had taken the visitors for either Hollanders or Frenchmen, the latter being particularly suspect because the previous Commander of Kamchatka had been murdered by a Hungarian Count who was believed to have been supported by the French government. Sergeant Surgutsky was reassured by Port's explanations of his visitors' intentions and relaxed his military precautions.

Coincident with this lessening of tension came a quickening of the season. Snow began to melt, greenery began to appear, and shore parties were sent to chop wood, tap birch trees for syrup, and gather nettles and other greens for the mess tables. The carpenters were set to work repairing the *Resolution*'s sheathing, which was "intirely eaten by the worms, to a perfect hony-comb."

Those who did not spend their time exploring and hunting wild duck soon discovered that the Russian community of Petrapavlovsk was engrafted upon a much depleted aboriginal population of Paleo-asiatic people called the Itelman or Kamchadal. The English had noticed the Kamchadal in Unalaska, where they acted as servants to the Russian fur traders. These people had been distinguishable from the Aleut, whom they in some ways resembled, by the fact that they did not perforate their lips (Plates 35, 36). Otherwise these two aboriginal peoples had seemed much alike, both serving their Russian masters much as though they were slaves. In Kamchatka, the Kamchadal were a subject population who provided goods and services to their overlords. Samwell reported that they were "superintended . . . in all their Towns by a Chief, a Countryman of their own, who is called Taionn; the Taionns are appointed by the Governor of Kamchatka who can displace them when he thinks proper, this Office is to preside over their own people but under the Controul of the Russians."

The Kamchadal dog-sleds were always at the service of the English after that first encounter, and the Chief or Taionn of the Kamchadals in Petrapavlovsk provided the ships with a daily supply of fresh fish. They spoke Russian as well as their own language and were baptized Christians. They dressed in a combination of traditional dog-skin clothing along with items of Russian and Chinese manufacture (Plates 56, 57). "Like all other People who are under the Yoke of Slavery," wrote Samwell, "The Kamtschadales are modest and submissive in their Behaviour to the Russians, & of late we learn that there has been no Disturbance among them but that they submit quietely to ye Government of their Masters."

From what may be gathered in the journal accounts, the English were preoccupied with the Russians during much of their stay at Petrapavlovsk, though King met a number of Kamchadals en route to Bolsheretsk, as did also Samwell in his travels. However, King's interests were focussed on Major Behm, and Samwell frequently sought out the com-

PLATE 54

A Man of Kamtschatka, [Kamchatka], Travelling in Winter.

pany of a "decent and venerable looking" priest named Vereshaggen, who lived some twenty miles away in the village of Paratunka. Captain Clerke remained confined to his cabin, making only one brief visit to the village on 12 May. Concern for the English Captain's declining health spread to the Russian community; Surgutsky provided him with beef and Vereshaggen daily sent a supply of fresh milk.

Lieutenant King's expedition to Bolsheretsk stretched to three weeks. King, Gore and Webber, accompanied by Port, a merchant-emissary, Fedositsch, and two cossacks, set out for the Avatcha River on the morning of 7 May 1779. They were poled up the river in Kamchadal *bats,* shallow river canoes which were the first boats to travel the river after the break-up of the ice. The following day they reached the *ostrog* (village) of Karatchin, where they were guests of the Taionn. He was "a very decent behav'd man," according to King, and the Kamchadal men and women were attentive and warmly welcoming. "Every one was drest out in their best," King wrote, "& the womens dress was very pretty & gay, compos'd of diff^t colourd Nankins, & some had part of their dresses made of a slight silk; their shifts were also silk, & the Married women had hansome Silk Handkerchiefs bound round their heads." Webber's portrait of "A Woman of Kamtschatka" (Plate 57) shows a particularly strong, persevering face, whereas in "A Man of Kamtschatka" (Plate 56) there is a powerful self-awareness behind the drawn eyebrows and taut mouth that suggests an arrested power. Of all Webber's portraits these are the most expressive. There are few comments from the journals to draw upon for interpretations of Kamchadal personalities, but certainly neither of these individuals could be termed meek or submissive-looking.

On the 9th, the party resumed their journey, this time by dog sledge. "This mode of travelling was so curious to us that we enjoy'd it prodigiously," remarked King. Driving required some skill, which meant that King and company remained passengers. King admired the great strength and agility of the Kamchadal drivers, but found that a sense of humour was necessary. "I had a very good natur'd Cossask who was however very unskillful," he recounted, "& we were upset every mile to the great mirth of the rest." Webber's "A Man of Kamtschatka, Travelling in Winter," (Plate 54) captures the heavy, sodden atmosphere of the thawing countryside. The static rendering of the dogs emphasizes the stillness but fails to convey the responsiveness and stability of the lightweight dog sledge that was so perfectly suited for icy terrain.

At dusk on 12 May the party finally arrived, fatigued, unkempt and quite unpresentable at "this Metropolis of Kamchatka," and to their great mortification saw that the town had turned out to receive them. They stopped short at a cottage on the outskirts of the *ostrog* to make themselves more presentable, then went to meet the Governor, who was waiting for them on the river's bank.

"I observd my Companions as awkward as myself

PLATE 55

The Inside of a Winter Habitation, in Kamtschatka [Kamchatka].

PLATE 56

Man of Kamtschatka [Kamchatka].

in making our first Salutations," confessed King, "bowings and scrapings being marks of good breeding that we had been now for 2½ years totally unaccustom'd to." The Major was truly a gentleman and his wife a lady of "good sence & breeding." The English guests were treated as ambassadors from the King of England; a residence was given over to them with domestic servants and guards placed at the door. Indeed, in this nondescript village of Bolsheretsk, situated in a remote Siberian wasteland (Plate 59), they discovered a truly genteel society. Major Behm refused to hear of payment for the bullocks and flour requested by Captain Clerke, insisting that his empress would not accept bills from people who were her allies and friends and who had come to her for assistance. Learning that the English sailors desperately wanted tobacco and were willing to pay any price, no matter how inflated, Behm presented his visitors with 400 pounds of tobacco to be given to the sailors. For Captain Clerke he sent butter, figs, rice and honey.

A party was given for them on their last evening and, "in the midst of the Wildest & most dreary country in the world," all the ladies of the village appeared elegantly dressed in Russian gowns while Mrs. Behm wore a strikingly rich silk dress from Europe. They danced the dances of "the Polite part of their empire," and King, Gore and Webber were filled with a nostalgia for their own society.

Major Behm decided to accompany his guests back to Petrapavlovsk, and on the morning of 16 May the entire town turned out to send them off with a procession of the village notables and a drum escort. People lined the streets, singing; the ladies were resplendent in silk cloaks lined with luxurious furs. A gun salute announced their departure and three hearty cheers sent them off in their heavily burdened canoes.

Behm's visit to the *Resolution* called for full honours to be accorded him. The marines were mustered and Captain Clerke presented Major Behm with some of his Polynesian "curiosities," these things being just about the only appropriate gifts for a man whose generosity was already beyond payment. Even the sailors were moved by Behm's gift of the tobacco and voluntarily offered to give up their grog in order to present the brandy to the Major. However, the Major declined this gift, saying that their journey ahead was to be a cold business. A few days later the ships were supplied with some fifteen thousand pounds of rye flour and twenty head of cattle. Fresh beef and a full allowance of bread became daily fare.

With a difficult and dangerous journey just ahead of them, Clerke determined to make a special request of Major Behm, who was preparing to leave his post in Kamchatka and return to St. Petersburg. Behm graciously agreed to take with him to St. Petersburg Captain Cook's and his own journals, a letter to the British Admiralty and a chart of their discoveries, and to despatch these to London.

It was mid-June, exactly on schedule, when Clerke took the *Resolution* and the *Discovery* out of Avatcha

PLATE 57

A Woman of Kamtschatka [Kamchatka].

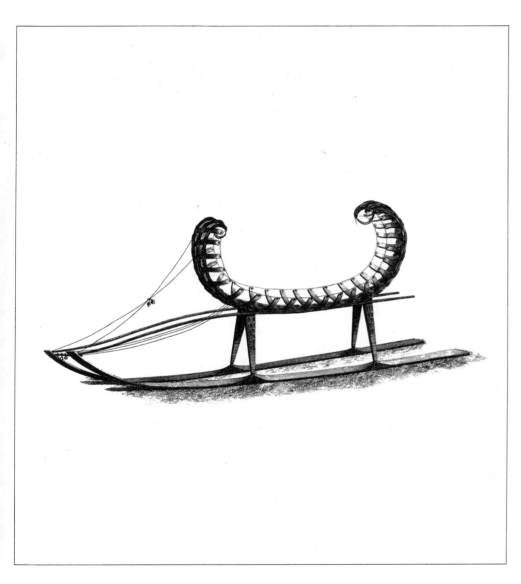

PLATE 58

A Sledge of Kamtschatka [Kamchatka].

Bay. As if to send them off with the grandest of fireworks, the Avachinskaya volcano thundered and spewed forth showers of cinders and dust on them. The sea, the air and the two boats were blanketed in ash, while all night the volcano grumbled and flashed behind them.

Soon the ships were once again into the sea-born fogs, pressing steadily north past St. Lawrence Island, where Anderson had been buried the previous year, then through Bering Strait. Within three weeks they faced the ice wall once again in almost the same latitude that they had encountered it in 1778. The ice rose above them some twenty feet out of the water, rough, uneven and unbroken as far as could be seen from the masthead. Clerke ran the ships west along its edge through drift ice that jarred against the bows, sometimes bringing them to a standstill. Once again the walruses, their only companions on this summit of the world, watched and brayed at them. Knowing full well that no amount of assiduous searching was going to bring them to a Northwest Passage, Clerke nevertheless doggedly persisted in running along the entire edge of the ice from east to west, from one continent to the other. When he was satisfied that he had seen with his own eyes the impenetrable barrier of ice that guarded the North Arctic Sea, he turned the expedition southward and headed back to Petrapavlovsk. Two days later, on 21 July 1779, his journal ended. He had only one month to live.

On the journey south the ships met up with the hunters of the northern ice (Plate 60). They sighted polar bears swimming and on the ice floes, and the sailors, insatiable hunters and carnivores themselves, one day gave chase in the jolly-boat and killed two of them, the largest bear weighing just under five hundred pounds. The flesh tasted somewhat fishy but was an improvement upon salt meat. King added a note that "in their maws were undigest'd pieces of seahorse [walrus], to those who could even eat Seahorse, this became a dainty food."

Just out of Avatcha Bay on 22 August 1779, Captain Clerke passed away. "Never was a decay, so melancholy & gradual," wrote King, whose admiration for Clerke's fortitude in commanding the expedition to the last was unstinting. Clerke had expressed a desire to be buried in Vereshaggen's church at Paratunka. King hoisted the *Resolution*'s colours at half-mast and brought the two ships to anchor before the town of Petrapavlovsk.

The "most dreary prospect" that had greeted them in April was in August totally transformed. "Every hill were most elegantly diversified with Trees, of different bloom, each having the greatest verdure conceivable, in short the whole face of the Country carrying a most butiful prospect." Webber's drawing of "A View of the Town and Harbour of St. Peter and St. Paul in Kamtschatka" (Plate 61) captures the warmth and peacefulness of Kamchatka in summer.

As soon as the ships were anchored, Sergeant Surgutsky boarded the *Resolution* to pay his respects to Clerke. On being informed of his death and his desire to be buried at Paratunka, Surgutsky once again looked dubious, referring them to Vereshaggen's authority. As it turned out, not only was it not the custom of the Greek church to bury their dead within sacred precincts but also the English discovered to their amazement that their Russian friends did not consider them Christians. Vereshaggen chose a spot near where a new church was to be built in Petrapavlovsk. As if to defy Vereshaggen's prejudice, Clerke's remains were committed "with all the military Honours due to his Rank & *according to the Ceremony of the Church of England*" (author's italics).

The English remained seven weeks in Petrapavlovsk, a stay which Gilbert described as tedious, though at least one man, John Holloway, a marine drummer on the *Discovery*, found romance. He deserted in order to remain with a Kamchadal woman, but was brought back, as were all those who had deserted under Cook's command. For the officers, their most interesting encounters were with the Russians. With Major Behm's departure a different style of colonial politics had emerged, this one characterized by petty power-mongering. A military officer appointed by Behm arrived to assume the command of Petrapavlovsk and promptly turned Surgutsky out of his home. A few days later Surgutsky received a severe flogging at this officer's hands on account of a disagreement. The English took Surgutsky's humiliation as an insult directed at them. When the new Governor of the area arrived from Bolsheretsk, the English interceded on Surgutsky's behalf; he was consequently reinstated and his superior officer sent back to his post in Ochotsk.

PLATE 59

A View of Bolcheretzkoi [Bolsheretsk], in Kamtschatka [Kamchatka].

PLATE 60

A White Bear.

PLATE 61

A View of the Town and Harbour of St. Peter and St. Paul, in Kamtschatka [Kamchatka].

In fact, Petrapavlovsk possessed all the ingredients for a fine dramatic Russian novel: characters such as Ivashkin, a Russian nobleman banished to Kamchatka on account of a youthful indiscretion with the empress; a murderous Hungarian Count who left the villagers suspicious of strangers; the gentlemanly Governor, Major Magnus von Behm, who was born in Livonia but was serving his empress in this remote corner of Empire; the priest Vereshaggen, whose mother was Kamchadal and father Russian.

When the expedition weighed anchor in Avatcha Bay, winter had already begun to strip the green from the trees and bushes, and the chill air hinted at snow. John Gore had taken over the command of the *Resolution* and King had succeeded him as captain of the *Discovery*. A hard year-long voyage lay before them, full of strenuous labour and the tedium of daily shipboard life.

Conclusion: Kamchatka to Plymouth

Cook's instructions for the Third Voyage had left open his choice of routes for the journey home. Captain Gore, in consultation with the other officers, decided to return via the Cape of Good Hope and, on the way, to trace the eastern coast of Japan and the little known Kurile Islands, which, according to the Russians, "were inhabited by people of a gigantic size, who were covered in hair." On 10 October 1779 the two ships sailed out on their homeward journey. Since they were now, as King wrote, "about to visit nations, our reception among whom might a good deal depend on the respectability of appearance," Gore ordered two of the *Resolution*'s guns to be taken out of the hold and mounted on deck.

The journey south, wrote Gilbert, "was the most disagreeable we had ever experienced; having a continual gale of wind, with very severe Squalls, Thunder, Lightning, and Rain, and an extraordinary high sea." Unable on this account to examine the Kurile Islands or Japan, they continued nearly south until Midshipman Trevenen sighted one of an unknown group of islands, Iwo Jima, whose volcanic activity covered the surrounding sea with powdery pumice. Gore named it Sulphur Island.

At last, on 4 December 1779, after narrowly missing being wrecked on the deadly Prata shoal in the China Sea, the ships anchored off Portuguese Macao, "to our inexpressable joy and satisfaction," wrote Gilbert, "having had no intiligence from Europe, for a space of three years: it being now exactly that time since we left the Cape of Good Hope." Before going ashore, however, the captains carried out Admiralty orders that the results of the expedition should be kept as secret as possible. All journals, charts and other papers concerning the voyage that had been kept by the officers and men were taken from them, "to prevent," wrote Gilbert, "any person publishing an account of our discoveries; but such as their lordships should appoint." Captain King stated that all documents were "given up with the cheerful compliance of all the officers." But Macao was alive with the news that the French as well as the Americans were at war with England, prompting Burney to make a minutely written copy of his journal on China paper, as a precaution against capture on the journey home.

Work began immediately to put the ships in a proper state of defence. A bower anchor from the *Resolution* was exchanged for four six-pound cannons from a British Indiaman then in port, bringing the *Resolution* up to sixteen guns and the *Discovery* up to ten. King set out for Canton, some seven miles away, to arrange for supplies, travelling first by brig and then, to his great enjoyment, by sampan. On his return, he brought the good news that the French and Americans, in their admiration for Cook and his contribution to knowledge, had exempted his ships from attack or capture. In return, Gore resolved to observe a strict neutrality.

At Macao and Canton the sea-otter skins that the crew had collected at Nootka Sound and along the coast of Alaska, and which they had used for bed coverings and clothes, were found to be highly valued by the Chinese. At Canton King obtained about eight hundred dollars for twenty skins which had belonged to Cook and Clerke. Some of the crew made an even greater profit. He wrote that "One of our seamen sold his stock, alone, for eight hundred dollars and a few prime skins, which were clean and had been well preserved were sold for one hundred and twenty each." In total, the crew sold their collection of furs for the enormous sum of two thousand pounds. The prospect of even greater wealth quickly drove away all thoughts of home from many minds. Some of the seamen were eager to return to the northwest coast at once. Two of the *Resolution*'s crew, in fact, deserted in the night with the ship's cutter, "seduced," King supposed, "by the prevailing notion of making a fortune." King, who himself later developed a plan for a surveying and fur-trading voyage, wrote, "The rage with which our seamen were possessed to return to Cook's River, and, by another cargo of skins, to make their fortune, at one time was not far short of mutiny...."

The trading between the Chinese at Macao and the crews of the *Resolution* and the *Discovery* produced some interesting and colourful fashions in the seamen's dress. The Chinese traded in goods as well as money, and the crew, whose European clothes had long since worn out, appeared, King noted, "mixed and eked out with the gaudiest silks and cottons of China."

On the voyage from Macao, the ships stopped at the island of Poulo Condore off the coast of South Vietnam, where they obtained supplies of fresh buffalo meat. They by-passed Singapore to sail south to the passage between Sumatra and Java at Batavia, stopping only to take on water at the island of Krakatau in the Sunda Strait, described by Gilbert as "the hotest and most unhealthy place in the world." They felt themselves lucky to pass through the strait without the death of a single man.

It was Gore's intention to now sail directly from Java to England, but in April urgently needed repairs to the ships after the passage across the Indian Ocean made it necessary to put in at Cape Town, which they left after a month on 9 May 1780. On the way, both ships celebrated the fourth anniversary of the beginning of the voyage with a double ration of grog.

It was to be several months more, however, before they could celebrate the end of the voyage. Contrary winds forced the ships north from the English Channel to a safe anchorage at Stromness Harbour in the Orkney Islands. King, with Bayly and Webber, travelled overland to London with the journals, sketches and charts, while Gore, who was taken ill, waited a month for favourable winds to carry the ships down the east coast to London. To the midshipmen, eager to see their families and to join the British war fleet in search of battle and prize money, the waiting was unbearable. At least one member enjoyed the stay, however; Sergeant Gibson of the marines, veteran of three voyages and the hero of at least one other island romance, got married. Sadly, just three days after leaving his bride to sail to London, he died and was buried at sea. Two weeks later, the ships were in the Thames.

They had been away for four years, two months and twenty-two days on the longest single exploring voyage in history, but there was little excitement to greet them. News of Cook's death had already reached England some ten months before, in the despatches sent overland by Clerke from Kamchatka. There was little else that they could tell the Admiralty about the main purpose of the voyage: they had found no Northwest Passage. Now they learned that the Pickersgill expedition to the eastern Arctic had ended in dismal failure some three years before.

Officers and men awaited promotion and dispersal to war duty. Gore, who had barely survived the voyage bringing the expedition home, was appointed to captain, Greenwich Hospital, the position vacated by Cook in 1776. He died in 1790. King was also made captain, and sailed the following year in charge of a five-hundred-ship convoy to the West Indies. Like Gore, he had returned from the Third Voyage in poor health and he died from tuberculosis in Nice in 1784 after finishing the official account of the voyage from where Cook had left it. James Burney was promoted to commander, and to captain in 1782. He fought against the French in the West Indies but came home in ill health and eventually retired as rear-admiral. Most of his later years were spent chronicling discoveries in the Pacific Northwest and northeast Asia. Trevenen, Vancouver and Riou were promoted to lieutenant, as was every mate and midshipman on the voyage who had served their time — some seventeen in all. Bligh, for some reason, was one of the few who did not receive immediate promotion. It was not until late in 1781, after some fighting service, that he received his commission. Molesworth Phillips was promoted to captain and married Burney's sister, to whom he had been introduced at the end of the voyage. When he died in 1832, aged seventy-five, he was the only surviving officer of the voyage.

Some of those who had served with Cook sailed together again. Samwell was appointed surgeon to King's ship the *Crocodile,* with Trevenen, who remained with King during his last days at Nice, as lieutenant. Several others went with them. Samwell described the pleasure they took in each other's company, writing in 1781 that "I am very agreeably situated in the *Crocodile* with Captn King and the rest of my old shipmates," adding that "we find no small satisfaction in talking over the eventful history of our voyage and are happy beyond Measure when any of our old Companions come to see us from other ships which they do as often as they can."

John Webber was hired to supervise the production of sixty-one engravings chosen by the Admiralty from his copious portfolio of drawings to illustrate the published account of the Third Voyage. The work was contracted out to several London engravers and took four years to complete. The success of the engravings led Webber to work up a number of his other sketches into engravings and aquatints, which he sold privately or published over the years until his death in 1793.

Several of Cook's men published their memories of the voyage. Lieutenant Rickman, Corporal Ledyard and William Ellis presented their versions of the voyage several years before the three-volume official account recorded by Cook and King appeared in 1784. Heinrich Zimmerman, on his return to Germany, wrote the first account to appear in Europe. Burney also produced a history of the voyage, and Samwell wrote an account of Cook's death with a sketch of his life and character. The narrative written by George Gilbert immediately on his return from the voyage remained unpublished. He died of smallpox in 1781.

These published records of the voyage and the great number of private journals and diaries kept by the crews of both ships indicate that the seamen and officers recognized from the outset the importance of their undertaking. Their writings remain as lasting tributes to Cook, and in them may be found the true estimate of Cook's character and ability. Samwell's description of Cook perhaps says it best:

Nature had endowed him with a mind vigorous and comprehensive, which in his riper years he had cultivated with care and industry. His general knowledge was extensive and various: in that of his own profession he was unequalled. With a clear judgment, strong masculine sense and the most determined resolution; with a genius peculiarly turned for enterprize, he pursued his object with unshaken perseverance: — vigilant and active in an eminent degree: cool and intrepid among dangers; patient and firm under difficulties and distress: fertile in expedients: great and original in all his designs, active and resolved in carrying them into execution. These qualities rendered him the animating spirit of the expedition: in every situation, he stood unrivalled and alone; on him all eyes were turned: he was our leading star—which at its setting, left us involved in darkness and despair. His constitution was strong, his mode of living temperate: . . .

He was a modest man, and rather bashful; of an agreeable lively conversation, sensible and intelligent. In his temper he was somewhat hasty, but of a disposition the most friendly, benevolent and humane. His person was above six feet high, and though a good looking man, he was plain both in address and appearance. His head was small, his hair which was a dark brown, he wore tied behind. His face was full of expression, his nose exceedingly well-shaped, his eyes, which were small and of a brown cast, were quick and piercing; his eyebrows prominent, which gave his countenance altogether an air of austerity.

He was beloved by his people, who looked up to him as a father, and obeyed his commands with alacrity. The confidence we placed in him was unremitting; our admiration of his great talents unbounded; our esteem for his good qualities affectionate and sincere.

In exploring unknown countries, the dangers he had to encounter were various and uncommon. On such occasions, he always displayed great presence of mind, and a steady perseverance in the pursuit of his object. The acquisition he has made to our knowledge of the globe is immense, besides improving the art of navigation and enriching the science of natural philosophy.

He was remarkably distinguished for the activity of his mind: it was that which enabled him to pay an unwearied attention to every object of the service. The strict œconomy he observed in the expenditure of the ship's stores, and the unremitting care he employed for the preservation of the health of his people, were the causes that enabled him to prosecute discoveries in remote parts of the globe for such a length of time as had been deemed impracticable by former navigators. The method he discovered for preserving the health of seamen in long voyages, will transmit his name to posterity as the friend and benefactor of mankind: the success which attended it, afforded this truly great man more satisfaction than the distinguished fame that attended his discoveries.

Bibliography

BEAGLEHOLE, J.C.

The Case of the Needless Death: Reconstructing the Scene. In Robin Winks, ed. *The Historian As Detective*. New York: Harper & Row, 1968.

The Life of Captain James Cook. Stanford: The University Press, 1974.

BIRKET-SMITH, KAJ

The Chugach Eskimo. Copenhagen: Nationalmuseets Skrifter, Ethnografisk Raekke, vol. VI, 1953.

BUCK, PETER

Arts and Crafts of Hawaii. Honolulu: Bishop Museum Special Publications no. 45, 1957.

CHARD, CHESTER H.

The Kamchadal: A Synthetic Sketch. *Kroeber Anthropological Society Papers*, no. 8:9, 1953, pp. 20-44.

COLE, DOUGLAS

Cook at Nootka — The Engraved Record. *Canadian Collector* III, May 1976, pp. 27-29.

COOK, JAMES

A Voyage to the Pacific Ocean in the years 1776, 1777, 1778, 1779, and 1780.... vol. I and II written by Captain J. Cook, vol. III by Captain J. King. ed. John Douglas. 3 vols. London, 1784.

A Voyage towards the South Pole and round the World...In the Years 1772, 1773, 1774, and 1775, 2 vols., London, 1777.

COOK'S JOURNALS I, II, III

The Journals of Captain James Cook on His Voyages of Discovery, ed. by J.C. Beaglehole. I, The Voyage of the Endeavour, 1768-1771, Cambridge, 1955. II, The Voyage of the Resolution and Adventure, 1772-1775, Cambridge, 1961. III, The Voyage of the Resolution and Discovery, 1776-1780. Cambridge, 1967.

CRANZ, DAVID

The History of Greenland.... London: printed for Brethren's Society for the Furtherance of the Gospel among the Heathen, 1767.

CURTIS, EDWARD

The North American Indian, vol. 11. New York: Johnson Reprint Corp., 1970.

DAVIDSON, J.W.
AND DERYCK SCARR, eds.

Pacific Islands Portraits. Canberra: Australian National University, 1970.

DOUGLAS, JAMES,
14TH EARL OF MORTON

Hints offered to the consideration of Captain Cooke, Mr Bankes, Doctor Solander and the other Gentlemen who go upon the Expedition on Board the Endeavour. Manuscript. Commonwealth National Library, Canberra, dated Chiswick Wednesday 10th August 1768.

DRUCKER, PHILIP

The Northern and Central Nootka Tribes. Washington: Smithsonian Institute, Bureau of American Ethnology Bulletin 144, 1951.

ELLIS, WILLIAM

An authentic Narrative of a Voyage performed by Captain Cook and Captain Clerke, in His Majesty's ships Resolution and Discovery...including A faithful Account of all their Discoveries, and the unfortunate Death of Captain Cook. 2 vols. London, 1782.

ELLIS, WILLIAM

Polynesian Researches. 2 vols. London: Fisher, Son & Jackson, 1829.

EMERSON, N.B.

Unwritten Literature of Hawaii. Washington: Government Printing Office, 1909.

FORCE, ROLAND
AND MARYANNE FORCE

Art and Artifacts of the 18th Century. Honolulu: Bishop Museum Press, 1968.

FORNANDER, ABRAHAM

An Account of the Polynesian Race. 2 vols. London: Kegan Paul, Trench, Trubner & Co., 1890.

GERBI, ANTONELLO

The Dispute of the New World. Pittsburgh: University of Pittsburgh Press, 1973.

GIFFORD, E.W.

Tongan Society. Honolulu: The Museum, Bishop Museum Bulletin 61, 1929.

GILBERT, GEORGE

Journal 1776-1780. Typewritten transcription. Special Collections Division, Main Library, University of British Columbia.

HALPERN, IDA

Nootka Indian Music from the Pacific North West Coast. Annotated. New York: Ethnic Folkways Library Album #FE 4524, 1974.

HAWKESWORTH, JOHN

An account of the Voyages undertaken...for making Discoveries in the Southern Hemisphere. 3 vols. London: W. Strahan, T. Cadell, 1773.

HRDLICKA, ALES

The Aleutian and Commander Islands. Philadelphia: The Wistar Institute, 1945.

HOWELLS, WILLIAM

The Pacific Islanders. London: Wiedenfeld and Nicholson, 1973.

JOCHELSON, WALDEMAR

History, Ethnology and Anthropology of the Aleut. The Netherlands: Anthropological Publications, 1966.

KAMAKAU, SAMUEL M.

The Works of the People of Old. ed. by Dorothy B. Barrere. Honolulu: Bishop Museum Press, 1976.

KOPPERT, VINCENT

Contributions to Clayoquot Ethnology. Washington: Catholic University of America Anthropological Series no. 1, 1930.

LEDYARD, JOHN

A Journal of Captain Cook's last Voyage to the Pacific Ocean, and in quest of a North-West Passage, between Asia and America.... Hartford, 1783.

LEVIN, M.G.
AND L.P. POTAPOV

The Peoples of Siberia. Chicago: University of Chicago Press, 1964.

LIND, JAMES

A Treatise of the Scurvy. Edinburgh, 1753.

MALO, DAVID

Hawaiian Antiquities. Translated by Nathaniel Emerson, 1898. Honolulu: The Museum, Bishop Museum Special Publications no. 21, 1951.

MARINER, WILLIAM

An Account of the Natives of the Tonga Islands.... Edinburgh: printed for Constable, 1827.

MAUDE, H.E.

Of Islands and Men. Melbourne: Oxford University Press, 1968.

MOZIÑO, JOSÉ MARIANO

Noticias de Nutka. ed. by Iris Wilson. Seattle: University of Washington Press, 1970.

NELSON, EDWARD W.

The Eskimo About Bering Strait. New York: Johnson Reprint Corp., 1971.

OLIVER, DOUGLAS A.

Ancient Tahitian Society. 3 vols. Honolulu: University of Hawaii Press, 1974.

SAHLINS, MARSHALL

Social Stratification in Polynesia. Seattle: University of Washington Press, 1958.

SAMWELL, DAVID

A Narrative of the Death of Captain James Cook. London: printed for G.G.J. and J. Robinson, 1786.

SMITH, BERNARD

European Vision and the South Pacific, 1768-1850. Oxford: Clarendon Press, 1960.

TRAVERS, ROBERT

The Tasmanians. Melbourne: Cassell Australia, 1968.

WILLIAMSON, ROBERT W.

Essays in Polynesian Ethnology. ed. by Ralph Piddington. Cambridge: The University Press, 1939.

ZIMMERMAN, HEINRICH

Zimmerman's Captain Cook. ed. by F. W. Howay. Toronto: The Ryerson Press, 1930.

Illustration Acknowledgements

For permission to reproduce colour paintings the publisher thanks:

The Alexander Turnbull Library, Wellington, New Zealand, for the coat-of-arms of James Cook.
The Mitchell Library, State Library of New South Wales, Sydney, Australia, for the watercolour "His Majesty's Sloop-of-War, *Resolution*" by Lt. Henry Roberts, R.N.
The National Portrait Gallery, London, England, for "Capt. James Cook, R.N., F.R.S." by John Webber.
The Trustees of the National Maritime Museum, London, England for "Moorea, Society Islands" by John Clevely; "Capt. James Cook, R.N., F.R.S." by Sir Nathaniel Dance; and "Poedooa, daughter of Oree..." by John Webber.

For permission to reproduce the illustrations in Part I the publisher thanks:

The Centennial Museum, Vancouver, B.C., for the engraving of Captain Cook's murder, page 29.
The Early Canadian Cartography Section, National Map Collection, Public Archives of Canada, Ottawa, Ontario, for "A Plan of the River St. Lawrence...," page 4, and the chart of Newfoundland, page 5.
Mr. and Mrs. J.E. Horvath, Vancouver, B.C., for the map of the Pacific Northwest, page 21.
The Public Archives of Canada for the drawing of Omai, page 28.
Special Collections Division, The Library, University of British Columbia, Vancouver, for Cook's plan of the defences of Quebec, page 4; Capt. Samuel Wallis's arrival in Tahiti, 1767, page 7; the report and diagrams of the transit of Venus, page 9; the engraving of the Endeavour River, page 12; the chart of the Southern Hemisphere, 1777, p. 16; the map of Tartary, page 18; and the map of Brobdingnag, page 19.
The Trustees of the National Maritime Museum, London, England, for the photograph of Hadley's octant, page 13; the photographs of K1 and K3, pages 14 and 15; the portrait of Joseph Banks, page 16; the portraits of Cook's officers, pages 22, 23, 24, 25, 26; and the portrait of John Webber, p. 27.

The engravings illustrating Part II are reproduced from a rare copy of Volume III, *A Voyage to the Pacific Ocean...* (1784) generously loaned by the Special Collections Division, The Library, University of British Columbia, Vancouver. For *Master Mariner* the engravings have been assigned numbers which, being consecutive, differ from the plate numbers of the 1784 edition.